Biomaterials for Surgical Operation

Shuko Suzuki · Yoshito Ikada

Biomaterials for Surgical Operation

Humana Press

Shuko Suzuki
Graduate School
Department of Medical Engineering
Nara Medical University
Shijo-cho 840, Kashihara-shi
634-8521 Nara
Japan
suzukis@naramed-u.ac.jp;
shukosuzuki@hotmail.com

Yoshito Ikada
Graduate School
Department of Medical Engineering
Nara Medical University
Shijo-cho 840, Kashihara-shi
634-8521 Nara
Japan
ikada@naramed-u.ac.jp;
yyikada@wd5.so-net.ne.jp

ISBN 978-1-61779-569-5 e-ISBN 978-1-61779-570-1
DOI 10.1007/978-1-61779-570-1
Springer New York Dordrecht Heidelberg London

Library of Congress Control Number: 2011942918

Printed on acid-free paper

Humana Press is part of Springer Science+Business Media (www.springer.com)

Preface

Recent surgical operations have been enhanced by using high-level technologies such as medical robotics and endoscopes for minimally invasive operations. However, modest technologies are still widely used as basic surgical tools. Generally, surgical operations involve tissue incision, excision, resection, suturing, ligation, hemostasis, sealing, adhesion prevention, etc. These procedures require instruments and devices made of metals or polymers. Metallic instruments used for general surgery include knives, needles, forceps, scalpels, retractors, distractors, etc, while commonly used polymeric devices are sutures, gauzes, hemostatic agents, sealants, tamponades, tubes, catheters, etc.

This work focuses on polymeric surgical devices, especially bioabsorbable devices used during surgical operations and this is mainly due to the tremendous amount of investigations designed to improve these devices that are still problematic due to their poor performance. *Biomaterials for Surgical Operation* consists of nine chapters, devoting the first three chapters to serve as an introduction to the following chapters. First, undesirable biological reactions that occur during or after surgery which justify the use of the devices are briefly summarized. In addition, concise description is given to describe the class of bioabsorbable polymers, as they are major players throughout the rest of the chapters. Bioabsorbable polymers are most suitable for surgical operations since their function is required only during or for a few days after surgery, and a second surgery for their retrieval is not necessary.

Chapter 4 presents an overview of biomaterials used to stop bleeding and air leakage. Although they are called surgical adhesives or sealants, they are not used to bind tissues together but to seal small holes during surgery. Although the most preferentially used sealant is fibrin glue, its sealing capability is not optimal for surgeons, in addition to the potential risk of virus infection. Therefore, a large number of studies have been conducted to develop improved and less expensive sealants than current fibrin glues. A recent trend of research on fibrin glue substitutes is described in this chapter.

Currently, adhesion between the serous membrane and the opposing tissue is still a serious problem during surgery. Various biomaterials have been studied to address this issue, and their recent accomplishments are overviewed in Chap. 5, focusing on bioabsorbable barrier membranes. Another problem associated with tissue adhesion is the difficulty in clinical evaluation of the performance of newly developed membranes due to the subsequent surgery that is essential for direct evaluation.

Chapter 6 deals with bioabsorbable devices for bone fixation. When the devices were first developed about 3 decades ago, they attracted much attention from orthopedic surgeons as the secondary surgery for their retrieval appeared unnecessary. However, clinical application of bioabsorbable devices is currently much less frequently realized than that of metallic fixation devices, probably because the bioabsorption rate and the mechanical strength of the polymeric devices are both lower than first expected. Nevertheless, in maxillofacial surgery, bioabsorbable polymers have been widely used as mini plates and screws.

Promotion of wound healing is eagerly desired by surgeons, and growth factors may lead to a promising method. At present, the successful clinical applications are limited because effective carriers for the sustained release of growth factors to the target tissues are not readily available. An exception is the use of bone morphogenetic proteins coupled with collagen sponge carriers. It is expected that numerous other growth factors will be clinically applied when improved carriers prepared from bioabsorbable polymers become available in the near future. Chapter 7 provides insight into growth factors to be used for the promotion of wound healing. A promising strategy is the application of a patient's platelet-rich plasma, as it provides powerful growth factors for wound healing in their autogenous form coupled with fibrinogen, which can act as a carrier hydrogel for the sustained release of the growth factors.

Chapter 8 describes bioabsorbable sutures that are indispensable for surgical operation. However, advancements in bioabsorbable suture materials are limited due to their longtime use and nearly optimized potential. This is in marked contrast with other absorbable biomaterials described in Chapters 4 through 7.

It will be apparent that we have not prepared an exhaustive survey of the literature, but have preferred to give a detailed analysis from a practical viewpoint of surgery. We hope that this presentation will prove to be a suitable introduction to a clinically important field of biomaterials. If this hope is fulfilled, our aim will have been accomplished.

Nara, Japan Yoshito Ikada

Contents

Chapter 1
Introduction

Abstract During surgical operations, a variety of biomaterials have been used in form of string, fabric, sheet, powder, plate, and liquid in addition to knife, forceps, and other metallic devices. Most of these biomaterials or devices for surgery are composed of polymers, mostly bioabsorbable polymers, as they are no longer needed after repair of the tissue injuries formed during operations. This introductory section overviews situations associated to surgery, where bioabsorbable devices are employed. This is followed by emphasis on fundamentals of biomaterial, which include minimal requirements, biocompatibility, bioabsorption, possible toxic substances, and clinical applications related to these biomaterials, especially focusing surgical sealants, anti-adhesion barriers, fixation devices, and wound healing promoters.

Surgical operations use operative manual and instrumental techniques on patients to repair injury or arrest disease. They consist of physical intervention in tissues involving cutting of tissues or closure of previously sustained wounds. An incision is made to access the surgical site. Minimally invasive surgery with endoscopes involves small outer incisions to insert miniaturized instruments within a body cavity or structure, while open surgical procedures or laparotomies require large incisions to access areas of interest. The approach to the surgical site may involve several layers of incision and dissection, as in abdominal surgery, where the incision must traverse skin, subcutaneous tissue, three layers of muscle, and then peritoneum. Removal of an organ, tumor, or other tissues is called excision, while partial removal of an organ or other bodily structure is called resection. For instance, when a segment of intestine is resected, the two remaining ends are sewn or stapled together.

Large blood vessels exposed at a tissue incision may be clamped by ligation to prevent bleeding, while retractors are used to expose the site or keep the incision open. When the blood vessels are too small to be clamped, bleeding is prevented by electric cauterization (fusing of a wound with extreme heat) or through the natural clotting process. If blood oozes from the cut surface of a resected organ, antibleeding sealants (adhesives) or hemostats will be applied to stop the flow of blood. Another type of liquid flow may occur from tissues such as the dura mater (cerebrospinal fluid)

S. Suzuki and Y. Ikada, *Biomaterials for Surgical Operation*,
DOI 10.1007/978-1-61779-570-1_1, © Springer Science+Business Media, LLC 2012

and pancreas (pancreatic juice). Rupture of a pulmonary bleb or bulla will result in natural air leakage. In addition, when a diseased part of the lung is cut, air leakage may occur from the incised cross section. For these tissues, sealants are very effective in preventing liquid or air leak. Fibrin-based hemostatic glues were first developed in the 1940s, while glue components, human fibrinogen, and thrombin are still very expensive and scarce. The adhesives and sealants with hemostatic properties are particularly useful in parenchymal surgery, mostly in liver, pancreas, and spleen surgery, above all in the case of partial resections and when it is necessary to favor the hemostasis of large injured areas, with massive bleeding [1].

Some injury may be unexpectedly produced during surgery by instruments, gloves, etc. In most cases, wounds formed during surgical operations spontaneously heal. However, wound healing will eventually lead to problems, resulting in incomplete repair, often with scar formation. In such cases, adhesions will occur between tissues in contact. Bacterial inflammation also triggers tissue adhesion. Postoperative adhesions are a significant problem after open surgery, although they are not recognized as the most frequent complication of surgery. The economic burden of adhesions following surgery is enormous, considering their annual cost, which exceeds $1 billion in the USA [2] alone, but remains unchanged despite advances in strategies to prevent them. One most effective method to prevent tissue adhesion is the use of barrier materials against adhesion. According to the study reported by Parker et al. [3], patients undergoing lower abdominal surgery (excluding appendectomy) had a 5% risk of readmission directly related to adhesions in the 5-year period following surgery. Appendectomy was associated with a lower rate of readmission (0.9%) but was responsible for over 7% of all lower abdominal surgery readmissions. Panproctocolectomy (15.4%), total colectomy (8.8%), and ileostomy (10.6%) were associated with the highest risk of adhesion-related admission. In a review article, Brochhausen et al. discussed the pathogenesis of adhesion formation based on the physiological role of serous membranes. Furthermore, they described several biomaterial approaches used in the prevention of peritoneal adhesion formation and perspectives for possible future developments [4].

Reunion of fractured bones is one type of wound healing. To immobilize reunited bones until complete reunion, surgeons use bone fixation devices mostly made from metals in the shapes of pins, nails, screws, and plates. As these metallic devices require secondary surgery for their removal, bioabsorbable fixation devices have attracted attention. Bioabsorption rate versus mechanical strength is a trade-off problem in use of these devices.

Wound healing involves complex processes recruiting a variety of cells, cytokines, and growth factors. However, normal tissue repair will be achieved if amounts of these components required for wound healing are topically provided in timely fashion. Drug delivery systems (DDS), which have already a half-century history of research and development, may be useful for this purpose, e.g., through the sustained release of growth factors to wounds. An interesting approach is the combined use of platelet-rich plasma (PRP) and fibrin gel since PRP contains a variety of growth factors that are indispensable for wound healing while fibrin gel can function as a carrier of platelet growth factors for extended periods of time. Both PRP and fibrin

Table 1.1 Medical applications of biomaterials

Usage	Examples
1. General medical treatment accessories (disposable products)	Syringe, catheter, suture, infusion set, blood bag, hemostatic agent, surgical adhesive, wound cover, bone fixation device, etc.
2. Blood purification (extracorporeal circulation)	Hemodialysis, oxygenation (cardiopulmonary bypass), plasma exchange, blood detoxification, etc.
3. Body contour repair	Maxillofacial prosthesis, breast implant, orthodontics, etc.
4. Replacement of diseased tissues and organs	Contact lens, dental implant, pacemaker, joint replacements, ligament augmentation, artificial organs, etc.
5. Drug delivery	Controlled release of drugs, drug targeting, etc.
6. Regenerative medicine (tissue engineering)	Scaffold for cells to regenerate tissues and organs

Table 1.2 Requirements for substances used as biomaterials

Requirements	Remarks
1. Nontoxic	No pyrogenicity, no hemolysis, no antigenicity, no chronic inflammation, no carcinogenicity
2. Functional (effective)	Replacement of defective organs and tissues, temporary assistance in medical treatments
3. Sterilizable	Autoclave, ethylene oxide gas, high-energy radiation
4. Biocompatible	Refer to Fig. 1.1

gel can be prepared in autologous form from the blood of patients before surgery and will be given to the patient during surgery. Several techniques for platelet concentrates are available, but their applications have been confusing because each method leads to different product with different biology and potential uses. Therefore, Ehrenfest et al. classified the different platelet concentrates into four categories depending on their leukocyte and fibrin content to elucidate successes and failures that have occurred so far, as well as providing an objective approach for the further development of these techniques [5].

Surgical operation is finalized with the closure of incised tissues. Medical devices for wound closure comprise resources that include nonbioabsorbable and bioabsorbable sutures, staples, tape, and adhesive compounds. Choosing the proper materials and wound closure techniques ensures optimal healing. In order to appreciate the essentials of wound closure, it is first important to understand the process of wound healing.

The above-mentioned materials are classified as biomaterials. Table 1.1 shows how and where biomaterials are applied in clinical medicine. As can be seen, they are used routinely in surgery, dental procedures, and drug delivery, and contribute to surgery by composing all or part of living structures which replace physiological functions, such as heart valves and hip joints. As demonstrated in Table 1.2, the minimal requirements of biomaterials for medical applications include nontoxicity, functionality (effectiveness), sterilizability, and biocompatibility. Although currently available biomaterials meet most of these requirements, they lack biocompatibility. This biocompatibility is the ability of a biomaterial to minimize foreign-body

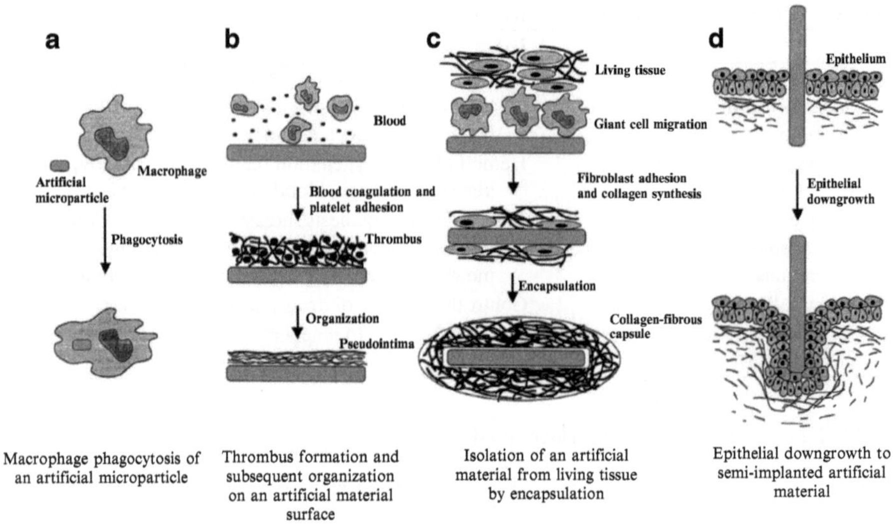

Fig. 1.1 Foreign-body-type reactions toward implanted biomaterials

Table 1.3 Biomaterial substances eliciting toxic response

Substances	Examples
1. Impurities	Unpolymerized monomer, catalyst, initiator fragment, mold-releasing agent, etc.
2. Additives	Antioxidant, UV absorber, plasticizer, pigment, etc.
3. Degradation (corrosion) products	Biodegradation fragments, wear debris, metal ions, etc.
4. Fragments of microorganisms	Endotoxin, antigenic biomacromolecule, cell debris, etc.
5. Material surface	Positive charge, sharp edge, etc.

reactions, as illustrated in Fig. 1.1. Biocompatibility thus cannot be defined in terms of material properties alone and involves a combination of material properties and functions for which their use is intended. Although a variety of materials have been explored for applications as medical devices, including polymers, metals, ceramics, and their composites, there is currently no biomaterial that is free of any significant foreign-body reactions when implanted in the body. The defensive mechanisms of the body respond to implants through immune rejection of them. The resulting fibrous encapsulation may lead to bacterial infection and eventually cause failure of the implant. Among biomaterials used as implants, sutures constitute the largest groups of materials, with an annual market exceeding $1.3 billion.

Biomaterials to be used inside the body must be approved by regulatory agencies. Regulatory authorities recognize different classes of medical devices fabricated from biomaterials, based on their complexity of design, characteristics of use, and their potential for harm if misused. The United States classifies the medical devices used for surgical operation as Class III devices. They are usually those that support or sustain human life, are of substantial importance in preventing impairment of human health, or prevent a potential, unreasonable risk of illness or injury. Possible toxic substances related to biomaterials are listed in Table 1.3.

This book covers polymeric medical devices that are essential for surgical operation, focusing in particular on bioabsorbable devices. Since the 1980s, sponge layers of collagen and chondroitin-6-sulfate, laminated with a silicone layer to a sheet, have been clinically used to repair skin wounds [6]. When this sheet is placed on wounds, the inner collagen layer turns into dermis-like connective tissue. This book does not consider these types of bioabsorbable devices since they are not commonly used in patients requiring surgery. Those who research the bioabsorbable devices to be described here have made great efforts to advance surgery, similar to those who research permanent devices such as artificial hearts.

References

1. Canonico S (2003) The use of human fibrin glue in the surgical operations. Acta Biomed 74: 21–25
2. van der Wal JB, Jeekel J (2007) The use of statins in postoperative adhesion prevention. Ann Surg 245:185–186
3. Parker MC, Wilson MS, Menzies D, Sunderland G, Clark DN, Knight AD, Crowe AM (2005) The SCAR-3 study: 5-year adhesion-related readmission risk following lower abdominal surgical procedures. Colorectal Dis 7:551–558
4. Brochhausen C, Schmitt VH, Rajab TK, Planck CN, Krämer B, Wallwiener M, Hierlemann H, Kirkpatrick CJ (2011) Intraperitoneal adhesions-an ongoing challenge between biomedical engineering and the life sciences. J Biomed Mater Res A 98:143–156
5. Ehrenfest DM, Rasmusson L, Albrektsson T (2009) Classification of platelet concentrates: from pure platelet-rich plasma (P-PRP) to leucocyte- and platelet-rich fibrin (L-PRF). Trends Biotechnol 27:158–16
6. Yannas IV, Burke JF, Orgill DP, Skrabut EM (1982) Wound tissue can utilize a polymeric template to synthesize a functional extension of skin. Science 215:174–176

Chapter 2
Biological Events Associated with Surgical Operation

Abstract During surgery, tissue injuries are created as a result of tissue incision and tissue or organ resection. The treatments often lead to bleeding at the site of injury. The body responds to this event initially with blood coagulation. This is followed by wound healing, a complex and dynamic process that involves a cascade of cellular reactions to restore tissue layer structures. The healing process can be divided into four phases: inflammation, proliferation, reepithelialization, and remodeling. Tissue adhesion is common after surgery due to severe injury. This chapter deals with the biological events that accompany surgical operations. Understanding this phenomenon is of great importance for the development of biomaterials used in surgical operations.

Blood Coagulation

The success of a surgical operation initiated by tissue incision depends on careful selection of the incision site while considering multiple factors. Among the most important is adequate exposure of the target tissue in approaching the operative field. In contrast to tissue incision, tissue or organ resection is often accompanied by bleeding (hemorrhage), especially when highly vascularized organs are subjected to resection. Blood vessels that bleed strongly on being injured or cut must be clamped before shock or death occurs. Bleeding from much smaller blood vessels may discontinue spontaneously through physiological clot formation.

Hemostasis, which is the process of blood clotting followed by clot dissolution and repair of injured tissue, is composed of four events that occur in a fixed order following loss of vascular integrity. As demonstrated in Fig. 2.1, the initial phase of the process is vascular constriction that limits the flow of blood to the area of injury. Next, platelets become activated by thrombin and aggregated at the site of injury, forming a temporary, loose platelet plug. To ensure stability of the initially loose platelet plug, a fibrin clot (network of insoluble fibrin molecules) forms and entraps the plug. Finally, blood clots are reorganized and resolved in order for normal blood

S. Suzuki and Y. Ikada, *Biomaterials for Surgical Operation*,
DOI 10.1007/978-1-61779-570-1_2, © Springer Science+Business Media, LLC 2012

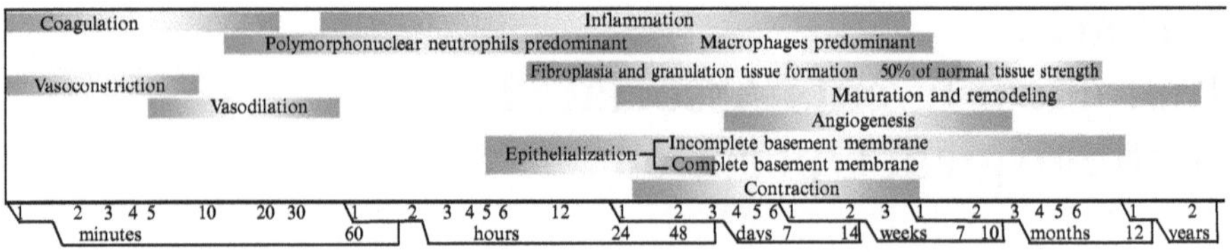

Fig. 2.1 Approximate time taken for different phases of wound healing

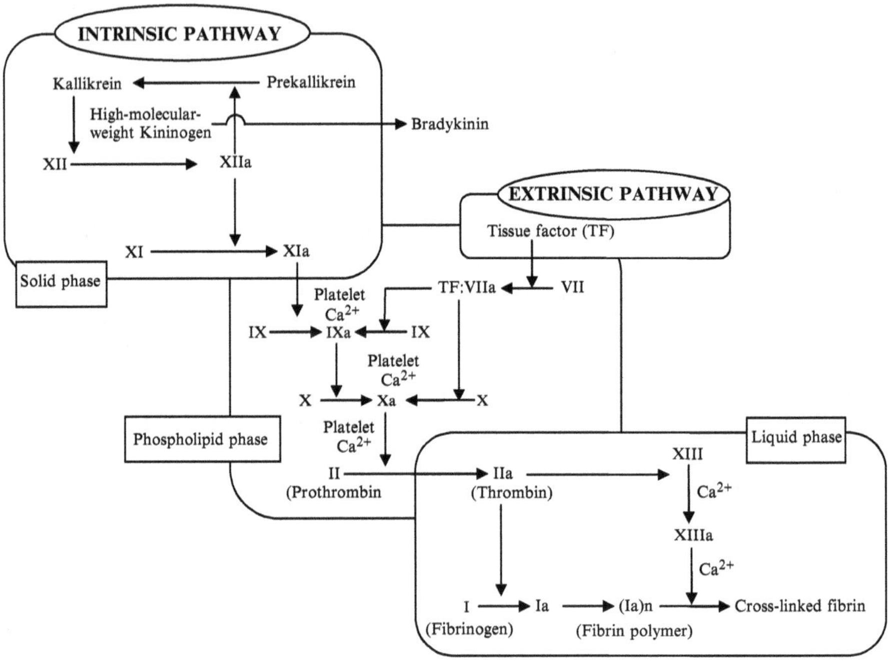

Fig. 2.2 Blood coagulation pathways

flow to resume following tissue repair. The dissolution of the clot (fibrinolysis) occurs through the action of the enzyme plasmin.

Both platelet aggregation and fibrin formation require the proteolytic enzyme thrombin. Clotting requires further calcium ions (Ca^{2+}) and about a dozen other protein clotting factors, as illustrated in Fig. 2.2. Most of them circulate in the blood as inactive precursors. They are activated by proteolytic cleavage becoming, in turn, active proteases (mostly serine proteases) for other factors in the clotting system.

Platelet Activation

In order for hemostasis to occur, platelets appear first and adhere to exposed collagen, release the contents of their granules, and aggregate, as shown in Fig. 2.3. Platelets are anuclear cell fragments produced from megakaryocytes. Blood normally

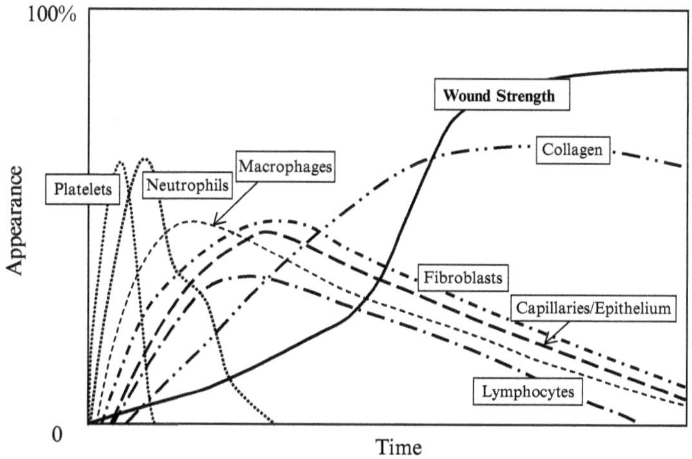

Fig. 2.3 Time-dependent cellular appearances during wound healing

contains 150,000–400,000 platelets per μl. The essential role of platelets is formation of a clot where blood vessels have been broken in addition to maintaining the integrity of adherens junctions that provide a tight seal between the endothelial cells that line blood vessels. When blood vessels are injured, fibrils of collagen in the extracellular matrix (ECM) are exposed. Platelets then begin to adhere to the collagen through the action of specific receptors for collagen present on their cell membrane. This adhesion is strengthened further by additional circulating von Willebrand factor (vWF) which forms additional links between platelet glycoproteins and the collagen fibrils. This adhesion activates platelets.

vWF is a complex multimeric glycoprotein that is produced by and stored in the α-granules of platelets. It is also synthesized by megakaryocytes and found in association with subendothelial connective tissue. The function of vWF is to act as a bridge between a specific glycoprotein complex on the surface of platelets (GPIb-GPIX-GPV) and collagen fibrils. In addition to its role as a bridge between platelets and exposed collagen on endothelial surfaces, vWF binds to and stabilizes coagulation factor VIII. Binding of factor VIII by vWF is required for normal survival of factor VIII in the circulation. The bound platelets also release messengers into the blood that perform additional functions. The released adenosine 5-diphosphate (ADP) and thromboxane A_2 recruit and activate still more platelets circulating in the blood to the injured site to enlarge the clot, while the released serotonin promotes constriction of the blood vessels to reduce bleeding.

The activation of platelets is induced by thrombin binding to specific receptors on the surface of platelets, thereby initiating a signal transduction cascade. The contents of stored granules are released from the activated platelets into the blood plasma and activate a Gq-linked protein receptor cascade, resulting in increased Ca^{2+} concentration in the platelet cytosol. The calcium activates protein kinase C, which, in turn, activates phospholipase A_2 (PLA_2). PLA_2 then modifies the integrin membrane glycoprotein, increasing its affinity to bind fibrinogen. Another enzyme activated by the released intracellular Ca^{2+} stores is myosin light chain kinase (MLCK). The activated MLCK phosphorylates the light chain of myosin, which then interacts with actin, resulting in altered platelet morphology (from spherical to stellate) and motility. Platelet activation is required for consequent aggregation to a platelet plug. Equally significant, however, is the role of activated surface phospholipids in activation of the coagulation cascade. Phospholipids are prominent components of platelet membranes, providing surfaces upon which the chemical reactions of coagulation can take place.

Extrinsic Clotting Cascade

Blood coagulation can be initiated by either of two distinct pathways: intrinsic and extrinsic. This is represented in Fig. 2.2. The intrinsic pathway requires only elements (clotting factors, Ca^{2+}, platelet surface, etc.) found within, or intrinsic to the vascular system, whereas the extrinsic pathway requires tissue factor (factor III, tissue thromboplastin), a substance "extrinsic to," or not normally circulating in, the blood vessel. This is therefore also known as the tissue factor (TF) pathway. The main role of the TF pathway is to generate a thrombin burst, a process by which thrombin, the most important constituent of the coagulation cascade in terms of feedback activation roles, is released. TF is released instantaneously when the vessel wall is ruptured. Thus, the extrinsic pathway is initiated upon vascular injury, which leads to exposure of TF, a subendothelial cell-surface glycoprotein that binds phospholipid. As shown in Fig. 2.2, the intrinsic and extrinsic pathways converge upon the activation of factor X to Xa. Factor Xa plays a role in the further activation of factor VII to VIIa. Active factor Xa hydrolyzes and activates prothrombin to thrombin. Thrombin can then activate factors XI, VIII, and V, furthering the cascade. Ultimately, the role of thrombin is to convert fibrinogen to fibrin and to activate factor XIII to XIIIa. Factor XIIIa (transglutaminase) forms covalent bonds between the soluble fibrin molecules, converting them into an insoluble network – the clot.

The clotting process includes several positive feedback loops which quickly magnify a tiny event into what may well be a lifesaving plug to stop bleeding. As indicated above, the clotting depends on a series of proteolytic reactions, in each of which an inactive precursor (zymogen) of a proteolytic enzyme is converted to an active enzyme. Because each step in the series is enzyme-catalyzed, and one enzyme molecule can theoretically catalyze the formation of a very large number of molecular products, this cascade has the capacity for enormous amplification. For example, one molecule of factor Xa can under ideal conditions generate about 1,000 thrombin molecules per minute. If two such reactions occur in sequence, the theoretical amplification is a millionfold per minute. Most proteins involved in clotting are plasma and not cellular components. The total protein concentration in normal plasma is ca. 70 mg/ml. The proteins devoted to clot formation account for less than 3 mg/ml, and of this, the bulk is fibrinogen. The remaining plasma clotting proteins

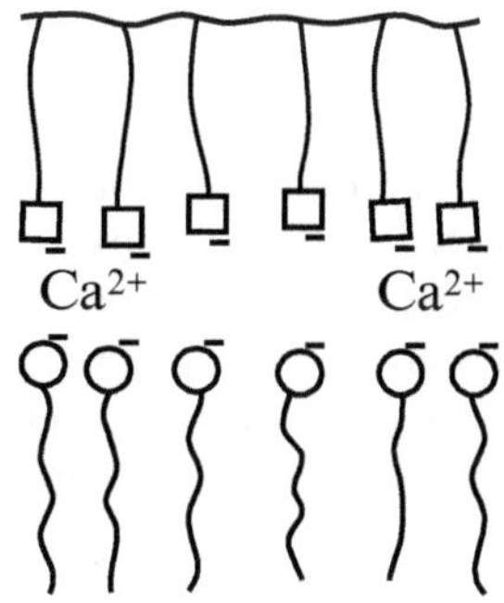

Fig. 2.4 Arrangement of negative "Gla" protein on the surface of the negative phospholipid layer of platelets with the help of positive Ca^{2+} ion

are present at much lower levels, ranging from prothrombin at about 120 μg/ml (1.6 μM), down to factors VII and VIII at less than 0.5 μg/ml (<10 nM).

TF is the protein that initiates normal coagulation, and appears to be absolutely required for hemostasis. Deficiency of TF has never been described. TF is an integral membrane protein with one transmembrane domain. It is not a zymogen, but a cofactor, and is normally expressed at very low, if at all, in the endothelial cells which line the blood vessels. Much richer in TF are cells that lie immediately beneath the endothelium, chiefly the fibroblasts and smooth muscle cells. The TF level in cells is under transcriptional control and can rapidly rise in response to several inflammatory and hormonal stimuli. Once the vessel wall is injured, TF in the subendothelial cells comes into contact with plasma proteins. Because TF is a transmembrane protein, it is unlikely to be released into the circulation unless there is massive tissue injury. It is generally accepted that it remains at the site of injury.

When TF comes into contact with the blood, it forms a complex with factor VII, TF:VII, though this complex has no proteolytic activity. To be active, a portion of factor VII must be activated to form the enzyme factor VIIa. Activated Xa leaves the TF-bearing membrane and is localized by its interaction with negative phospholipid, if it is available, as shown in Fig. 2.4. No normal cells have significant negative phospholipid head groups in the outer membrane leaflet. However, when cells (platelets, monocytes, etc.) are stimulated or injured, negative phospholipid appears on the outer leaflet of the phospholipid bilayer. Negative phospholipid binds Ca^{2+}, while the interaction of clotting proteins with negative phospholipid plays a role in the posttranslational carboxylation of glutamic acid residues in vitamin-K-dependent proteins. The immediate NH_2-terminal regions of these proteins are very rich in the modified amino acid–carboxyglutamic acid (Gla), termed the Gla domain. Ca^{2+} mediates the binding of complexes via the Gla domain of factor Xa to the phospholipid surfaces expressed by platelets, as well as procoagulant microparticles or microvesicles shed from them.

Thrombin has a large array of functions. Its primary role is the conversion of fibrinogen to fibrin, the building block of hemostatic plugs. In addition, it activates factor VIII, which forms covalent bonds that cross-link the fibrin polymers derived from activated monomers.

Wound Healing

Surgical operation inadvertently injures not only blood vessels but also other tissues, resulting in wound formation. Normally, wounds heal or repair themselves after injury. If the wound healing is delayed, tissue adhesion may occur. Postoperative tissue adhesion is a serious complication of surgery. To prevent adhesion, it is necessary to understand the mechanism of clotting. Indeed, great progress has been made in understanding the processes of wound clotting, but it involved principally skin wound healing. Wounds causing tissue adhesion after surgical operation appear to be related to the serous membrane covering viscera (internal organs) and their opposing peritoneum, pleura, or pericardium. The description below is based on cutaneous wound healing, since little is known concerning the wound healing associated with the serous membranes comprising the serosa.

Healing of the wounds created by incision by a scalpel, the trauma resulting from bullets, or the tissue death caused by conditions such as appendicitis all follow a similar and predictable process. The classic model of wound healing is divided into three sequential phases: the inflammatory, the proliferative, and the remodeling phases. Upon injury, a set of complex biochemical events takes place in overlapping fashion in a closely orchestrated cascade to repair the injury. The inflammatory phase is characterized by hemostasis (clot formation) and inflammation. Collagen exposed during wound formation activates the coagulation cascade, initiating the inflammatory phase. In the inflammatory phase, bacteria and debris are phagocytosed and removed. After injury to tissue occurs, the cell membranes, injured in wound formation, release factors that cause the migration and division of cells involved in the proliferative phase. As represented in Fig. 2.5, this phase is characterized by angiogenesis, collagen deposition, granulation tissue formation, epithelialization, and wound contraction. In angiogenesis, new blood vessels are formed from vascular endothelial cells.

Inflammatory Phase

Within the first 6–8 h following vasoconstriction and fibrin formation, the next step of the healing process is underway, with polymorphonuclear neutrophils (PMNs) engorging the wound. Fibrin provides the structural support for cellular constituents of inflammation. The inflammatory phase is initiated by the release of various soluble factors (including chemokines and cytokines) to attract cells that phagocytose debris, bacteria, and injured tissue. Within an hour of wound formation, PMNs arrive at the wound site and become the predominant cells in the wound for the first 2 days after the injury occurs, with especially high numbers of them present on the second day. Platelets release platelet-derived growth factor (PDGF) and transforming growth factor-beta (TGF-β) from their α-granules to attract PMNs and macrophages. Macrophages are the most important mediator-releasing cells, replacing PMNs as the predominant cells in the wound by 2 days after injury. Attracted to the

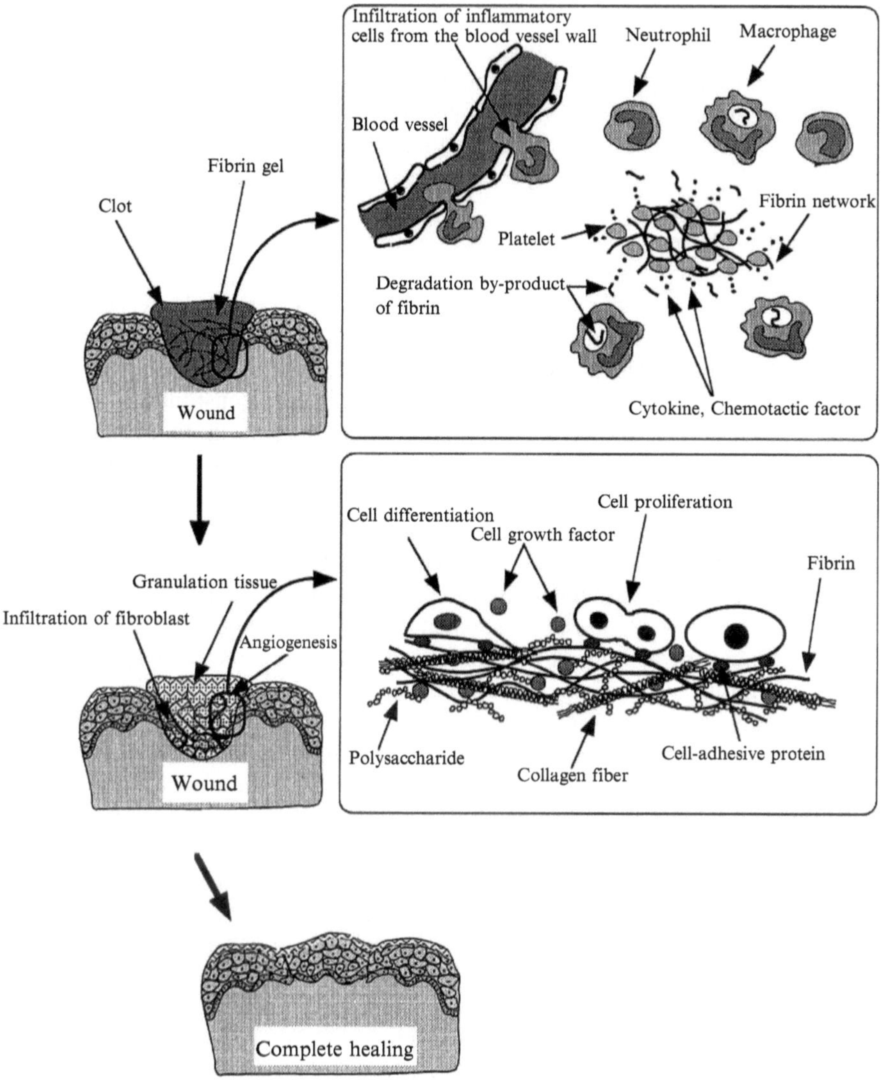

Fig. 2.5 Processes involved in wound healing

wound site by growth factors secreted by platelets and other cells, monocytes from the blood stream enter the area through blood vessel walls. Numbers of monocytes in the wound peak 1–1.5 days after the injury occurs. Once they are in the wound site, monocytes mature into macrophages. They orchestrate the multiplication of endothelial cells in tandem, with the sprouting of new blood vessels and duplication of smooth muscle cells. Many factors influencing the wound healing process are secreted by macrophages, especially during the third and fourth days after wound formation. These factors attract cells involved in the proliferation stage of healing to the area, although they may limit the contraction phase. Macrophages are stimulated by the low oxygen content of their surround to produce factors that induce and speed up angiogenesis. They also stimulate the cells that reepithelize the wound, create granulation tissue, and lay down new ECM. By secreting these factors, macrophages contribute to advancing the healing process into the next phase.

Proliferative Phase

About 2 or 3 days after the wound appears, fibroblasts begin to be recruited to the wound site, marking the onset of the proliferative phase even before the inflammatory phase has ended. The proliferative phase consists of different subphases. These subphases do not occur in discrete time frames and instead constitute an overall and ongoing process. The subphases are angiogenesis, fibroplasia, matrix deposition, and reepithelialization. During days 5–7, fibroblasts migrate into the wound, laying down new collagen of subtypes I and III. The most important cell in this phase is the fibroblast. Fibroblasts peak approximately day 7 from injury and are responsible for initiating angiogenesis, collagen formation, and reepithelialization.

Also termed as neovascularization, the process of angiogenesis occurs concurrently with fibroblast proliferation when endothelial cells migrate to the area of the wound. Because the activity of fibroblasts and epithelial cells requires oxygen and nutrients, angiogenesis is imperative for other stages in wound healing, including epidermal and fibroblast migration. Angiogenesis is the product of parent vessel offshoots. The formation of new vasculature requires ECM and basement membrane degradation followed by migration, mitosis, and maturation of endothelial cells. Basic fibroblast growth factor (bFGF) and vascular endothelial growth factor (VEGF) are believed to modulate angiogenesis.

One of the key roles of fibroblasts is the production of collagen. Collagen deposition is important because it increases wound strength. In addition, cells involved in inflammation, neovascularization, and connective tissue construction attach, grow, and differentiate into the collagen matrix laid down by fibroblasts. Type III collagen and fibronectin are generally beginning to be produced in appreciable amounts somewhere between about 10 h and 3 days after wound formation, mainly depending on wound size. Their deposition peaks at 1–3 weeks. They are the predominant tensile substances until the later phase of maturation, in which they are replaced by strong type I collagen. Granulation tissue, formed in the later phase, is particularly important in wound healing. When collagen synthesis and breakdown become equal in rate of occurrence, the next phase of wound healing has begun.

Reepithelialization Phase

The formation of granulation tissue in an open wound enables the reepithelialization phase to take place, as epithelial cells migrate across the new tissue to form a barrier between the wound and the environment. Basal keratinocytes from the wound edges and epidermal appendages such as hair follicles are the cells principally responsible for this phase of wound healing. They advance in a sheet across the wound site and proliferate at its edges, ceasing movement when they meet in the middle of the wound.

Remodeling Phase

After the third week, the wound undergoes constant alteration, known as remodeling, which can last for years after the initial injury has appeared. Increased collagen production and breakdown continue for 6 months to 1 year after injury. When the levels of collagen production and degradation equalize, the maturation phase of tissue repair begins. The initial type III collagen is replaced by type I collagen until a type I:type III ratio of 4:1 is reached, equal to that in normal skin. Originally disorganized collagen fibers are rearranged, cross-linked, and aligned along lines of tension.

Contraction of the wound is an ongoing process resulting in part from the proliferation of specialized fibroblasts termed myofibroblasts, which resemble contractile smooth muscle cells. Maximal tensile strength of the wound is achieved by the 12th week, and the ultimate scar has only 80% of the tensile strength of the original skin that it has replaced. Vascularity decreases, resulting in a less hyperemic and more cosmetically appealing wound as this phase progresses. Since activity at the wound site is reduced, the scar loses its red appearance as blood vessels that are no longer needed are decreased by apoptosis.

The timetable for wound response can be quite variable. Chronic wounds can stall in the inflammatory phase because of poor perfusion, poor nutrition, or a myriad of other factors, causing excessive buildup of exudates in the wound base. These phases can also be blunted, as in the fetus, which has a decreased inflammatory period and heals without scar. Thus, understanding the growth factors involved in fetal healing may lead to novel methods of treatment associated with less scarring.

Tissue Adhesion

Visceral tissue adhesion very often occurs after surgery, especially in abdominal, gynecologic, and thoracic surgery. This seems to occur as a result of injury produced during surgery followed by clot formation. Clotting promotes wound healing, but works negatively for tissue adhesion. Tissue adhesion probably will not occur if the surgical injury is not severe, and healing can occur without it, similar to the skin wound healing that occurs without scar formation. However, there is a substantial difference in repair processes between skin wounds and the visceral tissue injuries produced during surgery. In contrast to skin wounds, which are mostly open to the outer environment (air), the injuries observed in the case of surgery are not exposed to air but are instead in direct contact with body fluids such as peritoneal liquid. The outer surface of the skin tissue is stratified squamous epithelium, while internal,

Stratum corneum
Stratum lucidum
Stratum granulosum
Stratum spinosum
Stratum basale

Layers of the epidermis

Serosa (simple squamous mesothelium)
Longitudinal muscle
Myenteric plexus
Circular muscle
Submucosal plexus
Submucosal
Mucosal

Layers of the enteric tissue

Fig. 2.6 Structures of the epidermis and enteric tissue

visceral organs are covered with serous membrane. Figure 2.6 compares the structure of the epidermis with that of the intestinal wall as an example of visceral tissue. Culture of keratinocytes, the major cells of the epidermis, must be performed at air-liquid interfaces to obtain the epidermis-like, multi-layered structure. The outermost surfaces of the pleura, pericardium, peritoneum, intestine, and uterus are covered with serous membrane consisting of mesothelial cells and a small amount of connective tissue supporting the simple, squamous mesothelium. In contrast to the outermost surface of the epidermis, the serous membrane is surrounded by aqueous media. The environments surrounding the visceral tissues/organs and the fetal skin are very much alike. As mentioned above, scar formation is not observed in the process of fetal wound healing.

The processes beginning with surgical injury and leading to tissue adhesion are not clearly known. Although the early phases of tissue adhesion may be similar to those of the wound response described above, the mechanism of visceral remesothelialization must differ from that of cutaneous reepithelialization. If remesothelialization ends normally, tissue adhesion will not occur. It is unclear whether tissue adhesion is correlated with scar formation in wound healing. The clearest function of the epidermis is physical and chemical protection against a variety of outer pathological stimuli, while the mesothelial layer may have other functions and be more vulnerable than the epidermis.

When surgery is performed after opening of the thoracic or abdominal wall, the surgical cavity is exposed to the air and the light present in the operating theater, reducing the humidity around the tissue/organ surface. On the other hand, surgeries performed with use of endoscopes are carried out in closed rather than open systems, and the high humidity of the thoracic and abdominal cavities can be retained without drying of tissue/organ surfaces. Endoscopic manipulation is known to be associated with less tissue adhesion than open surgery. Most likely, unrecognizable, small injuries occur on the drier mesothelial surface due even to tiny mechanical stimuli such as contact with surgical forceps and gloves. This may account for the tissue adhesion that occurs during open surgery even when only one side of tissue/organ surfaces has been recognizably injured as lines of anastomosis.

It should be pointed out that tissue adhesion often occurs between injured serous membrane and omentum as a result of wrapping of the injured region by omentum. This omental wrapping may play a role in repair of injured tissue. Adhesion may also occur between two different surfaces of intestine, though the process involved in this is unclear.

Acceleration of Wound Healing and Tissue Regeneration

A significant contribution to surgery would be made if an additional, simple treatment could be applied to patients before the end of surgery to promote wound healing or regenerate diseased tissue. One widely accepted means of this is the supply of growth factors, while another is transplantation of stem cells. DDS, which emerged about half a century ago, involves the sustained delivery of growth factors. Bolus injection of growth factors in aqueous solution is only minimally efficacious, since the administered drug quickly diffuses away from the site of injection. To improve this simple means of delivery, the modern form of DDS makes use of drug carriers that facilitate the sustained release of drugs at target tissues. For example, collagen sponges have been clinically used in the United States as slow-release carriers of bone morphogenetic proteins (BMPs). Major obstacles to use of this type of DDS for delivery of growth factors include the high cost of these proteins in addition to the poor accessibility of growth factors to tissue.

This problem can be avoided if we use the patient's platelet that contains a variety of growth factors effective in wound healing, such as TGF-β, PDGF, and VEGF. Just prior to surgical operation, a small amount of blood can be collected from the patient and then subjected to preparation of PRP from the blood by centrifugation. The PRP preparations may sufficiently contain the growth factors required for enhanced wound healing or tissue regeneration, but most of the growth factor molecules will be diffused out without displaying their own biological potential, unless an adequate carrier is employed. Fortunately, a carrier can be prepared from the patient's blood as it contains the fibrinogen which is a good resource for carrier hydrogel. An addition of patient's thrombin to fibrinogen solution leads to hydrogel formation. Trapping the platelets into the fibrin hydrogel, followed by implanting the platelet-fibrin gel into the patient during the surgical operation, the rate of wound healing and tissue regeneration will be enhanced. The key factor for this strategy is how effectively the growth factors can be entrapped in the carrier gel and released to the target tissue in terms of optimal time and site.

Chapter 3
Bioabsorbable Polymers

Abstract In surgical operations, materials are required as a support only during surgery and for some length of time after surgery. Therefore, bioabsorbable polymeric materials are preferred over nonbioabsorbable polymers as they do not require a secondary operation for removal. While long-term biocompatibility is not necessary for these materials, it is important for the bioabsorption rate to be matched with the rate of tissue regeneration. A wide variety of natural and synthetic polymers with various bioabsorption rates have been explored. Natural polymers, including proteins and polysaccharides, degrade by enzymatic hydrolysis, whereas synthetic polymers, mainly poly(α-aliphatic ester)s, do not require enzymes for hydrolysis. This chapter discusses bioabsorbable polymers commonly used in surgical operations.

Introduction

Polymeric biomaterials can be divided into two types, nonbioabsorbable and bioabsorbable. Here, the terms "absorbable," "resorbable," and "bioresorbable" are considered to have the same meaning. Nonbioabsorbable polymers have been used as key materials for artificial organs, implants, and disposable medical devices. Bioabsorbable polymers are clearly not useful as major components of permanent devices since bioabsorption and degradation of materials in these applications have negative effects almost identical to the deterioration of materials, an undesirable characteristic of many biomaterials designed for permanent use. The widespread clinical use of silicone, poly(ethylene terephthalate)(PET), polyethylene, polytetrafluoroethylene (PTFE), and poly(methyl methacrylate)(PMMA) as indispensable components of artificial organs and tissues is due to their excellent chemical stability (nondegradability) in the body. Problems arise if these materials undergo detectable degradation in the body since by-products of degradation may evoke untoward reactions in the body. If these stable polymers exhibit deterioration over time, this may not be due to hydrolysis but to attack by active oxygen species generated in the body as a result of inflammation.

S. Suzuki and Y. Ikada, *Biomaterials for Surgical Operation*,
DOI 10.1007/978-1-61779-570-1_3, © Springer Science+Business Media, LLC 2012

Table 3.1 Clinical applications of bioabsorbable polymers

Function	Purpose	Examples
Binding	Fixation	Fractured bone fixation
	Adhesion	Adhesion of tissue pieces
Closure	Suturing	Tissue approximation and vascular and intestinal anastomosis
	Covering	Wound cover
	Sealing	Topical hemostasis and air leakage
	Occlusion	Vascular embolization
Separation	Isolation	Organ protection
	Contact inhibition	Adhesion prevention
Scaffold	Cellular proliferation	Tissue and organ regeneration
	Tissue guide	Nerve reunion
Capsulation	Controlled drug delivery system	Sustained drug release

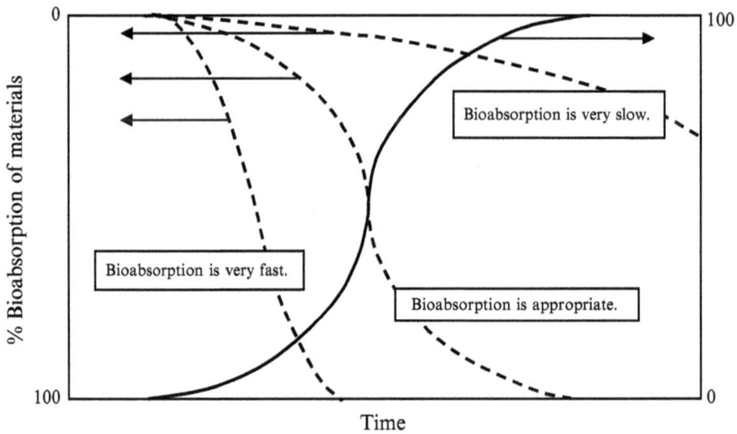

Fig. 3.1 Bioabsorption of materials during tissue recovery

Foreign-body reactions will continue to emerge as long as implanted permanent polymer remains in the body. Bioabsorbable polymers also elicit foreign-body reactions, which, however, cease to appear when the implanted materials have been completely absorbed into the body, together with the by-products of biodegradation such as oligomers and monomers. Bioabsorbable polymers have thus attracted much attention since they need not exhibit long-term biocompatibility. Applications of bioabsorbable polymers in medicine are briefly listed in Table 3.1. These materials remain in the body while functional, but disappear via bioabsorption thereafter. As a consequence of this, secondary surgery is not required to remove them after defects have been fully repaired. Despite these advantages, bioabsorbable polymers are not extensively used in current medical practice, principally for the following reasons. First, although the bioabsorption rate must match the rate of tissue regeneration, as shown in Fig. 3.1, it is difficult to meet this condition. If the bioabsorption of a

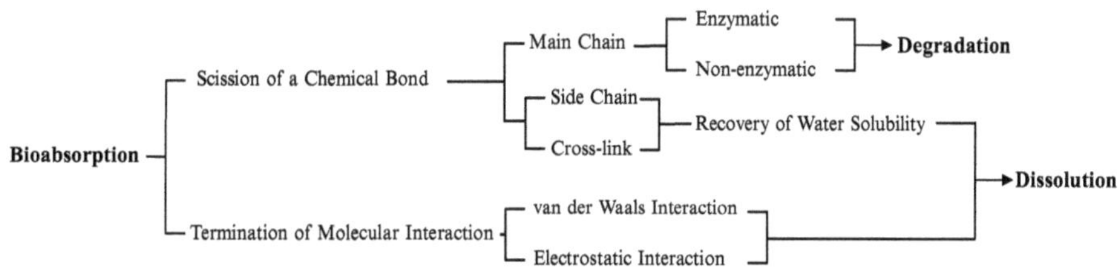

Fig. 3.2 Two modes of bioabsorption for polymers

material is faster than tissue regeneration, defects will recur. On the other hand, if the bioabsorption of a material is too slow, the material left behind may interfere with tissue physiology. Matching of the bioabsorption rate with the rate of healing of target tissue is one of the major challenges in designing bioabsorbable implants. The general criteria for selecting a bioabsorbable polymer for use as a biomaterial are matching of mechanical properties and the time required for bioabsorption with the needs of the application. This is further complicated by the fact that the rate of healing process is dependent on the patient as well as the extent of tissue injury.

The second reason is that these materials release degraded substances that may have toxic effects. The toxicity of biomaterials is mostly due to their low molecular weight, which enables them to reach many tissues. They include unpolymerized monomers, additives, fragments of polymerization initiators, and catalysts, as previously shown in Table 1.3. Currently, use of these compounds is regulated. Since degraded substances are constantly released in the body in the case of bioabsorbable materials, the safety of their metabolites must be carefully tested.

Furthermore, great care must be taken during the manufacturing and storage of hydrolysable polymers to prevent degradation of them due to exposure to water. Sterilization of these materials is another concern. For example, since poly (α-hydroxyacid)s are sensitive to radiation, sterilization is usually performed with ethylene oxide gas, which requires extensive degassing to remove residual traces of it.

Bioabsorbable polymers have several advantages over nonbioabsorbable, permanent polymers. Bioabsorbable polymers are very useful, particularly in surgery as a supporting material, since such support will, after some length of time, no longer be required postoperatively and since it enables healing of tissue injury produced during surgery. Bioabsorption of a polymeric implant does not require surgical intervention for removal at the end of its functional life, eliminating the need for secondary surgery.

Important criteria to consider when selecting a bioabsorbable polymer for biomedical applications are (1) mechanical properties, which must match the application and remain sufficiently strong until the surrounding tissue has healed, with degradation times matched to the times required; (2) lack of toxic responses with high metabolizability in the body after completion of function; and (3) ease of processing to the final product form with an acceptable shelf life and easy sterilization.

There are two groups of bioabsorbable polymers, one biodegradable and the other biosoluble. Figure 3.2 shows the molecular mechanism of bioabsorption of polymer. Biodegradable does not indicate degradability by enzymatic hydrolysis.

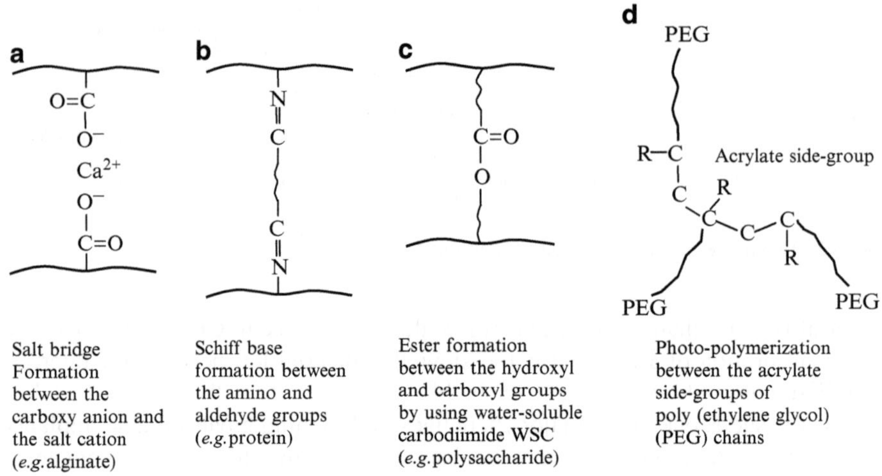

Fig. 3.3 Representative cross-linking reactions that take place under mild conditions

Physicochemical hydrolysis of a polymer at pH 7 without the contribution of enzymes also can result in degradation, followed by bioabsorption. Well-known polymers that undergo degradation through this mechanism include poly(glycolic acid)(PGA) and poly(lactic acid)(PLA). Both of these polyesters are bioabsorbed without the aid of esterases and lipases. On the other hand, biosoluble polymers are bioabsorbed without degradation as a result of their dissolution in the aqueous media of the body. This indicates that the original polymers of this subgroup should be quite soluble in water at pH 7 and 37°C. These water-soluble, nondegradable polymers can be used as implantable biomaterials if temporary, nonpermanent cross-linking can be introduced. An example is alginate, which is soluble in water but becomes insoluble after reaction with Ca^{2+} to form a salt bridge. Representative cross-linking reactions that take place under mild conditions are depicted in Fig. 3.3. The salt bridge formed with Ca^{2+} will be removed if the binding Ca^{2+} is replaced by that with Na^{+}. The cross-linking reactions shown in Fig. 3.3 can also be applied to degradable polymers, if they are soluble in water, such as proteins, as will be described below.

The following introductory description is limited to the bioabsorbable polymers that have been relatively widely used for medical purposes.

Biodegradable Polymers

Both natural and synthetic polymers are included among biodegradable polymers, as listed in Table 3.2. Table 3.3 displays the hydrolyzable bonds in the degradable polymers. The natural polymers include proteins and polysaccharides, while most

Table 3.2 Natural and synthetic biodegradable polymers

Natural polymers (enzymatic degradation)	Synthetic polymers (nonenzymatic degradation)
1. Proteins e.g., Collagen, gelatin, and fibrin	1. Polyesters e.g., Poly(glycolic acid) (PGA), poly(lactic acid) (PLA), glycolic acid/lactic acid (GA/LA) copolymer, GA/ε-caprolactone (CL) copolymer, LA/CL copolymer, poly(*p*-dioxanone) (PDS), and trimethylene carbonate/GA copolymer
2. Polysaccharides e.g., Starch, hyaluronate, chitin, chitosan, oxidized cellulose, and alginate	2. Polycyanoacrylates e.g., Isobutyl cyanoacrylate polymer
3. Polyesters e.g., Poly(β-hydroxybutyrate) (PHB)[a]	3. Polypeptides[b] e.g., Poly(L-glutamic acid) and poly(L-lysine)
4. Nucleic acids e.g., Deoxyribonucleic acid, ribonucleic acid	4. Others e.g., Polyanhydrides, poly(ortho esters), polyphosphazenes, and specific polyurethanes

[a]Nonenzymatic degradation
[b]Enzymatic degradation

Table 3.3 Hydrolyzable bonds in bioabsorbable macromolecules and their representative polymers

Bond	Chemical structure	Representative polymers
Ester	$-\underset{\underset{O}{\Vert}}{C}-O-$	PGA, PLA, Poly(malic acid)
Peptide	$-\underset{\underset{H}{\vert}}{N}-\underset{\underset{O}{\Vert}}{C}-$	Collagen, fibrin, synthetic polypeptides
Glycoside	$\begin{matrix}-O\\-C\end{matrix}>C-O-C<\begin{matrix}C-\\O-\end{matrix}$	HA, alginate, starch, chitin, chitosan
Phosphate	$-\overset{\overset{O}{\Vert}}{\underset{\underset{O}{\Vert}}{P}}-O-$	Nucleic acid
Anhydride	$-\underset{\underset{O}{\Vert}}{C}-O-\underset{\underset{O}{\Vert}}{C}-$	Poly(bis-*p*-carboxy phenoxy propane anhydride) $\mathbf{-\!\!(\!-CO-C_6H_4-O\,(\!-CH_2\!-\!)_3\,O-C_6H_4-\underset{\underset{O}{\Vert}}{C}-O-\!)_n}$
Carbonate	$-O-\underset{\underset{O}{\Vert}}{C}-O-$	Poly(ethylene carbonate) $\mathbf{-\!\!(\!-CH_2-CH_2-O-\underset{\underset{O}{\Vert}}{C}-O-\!)_n}$
Orthoester	$\begin{matrix}-C\\-O\end{matrix}>C<\begin{matrix}O-\\O-\end{matrix}$	
Carbon–carbon	$-CH_2-\overset{\overset{CN}{\vert}}{\underset{\underset{\underset{\underset{O-}{\vert}}{C=O}}{\vert}}{C}}-$	Poly(isobutyl cyanoacrylate
Phosphazene	$-\overset{\vert}{\underset{\vert}{P}}=N-$	Polydiaminophosphazene

of the synthetic biodegradable polymers are predominantly poly(α-aliphatic ester)s. Natural polymers always require enzymes for their hydrolytic degradation, while no enzymes are required for the hydrolysis of synthetic polymers.

Natural Polymers

The origin of naturally occurring polymers is human-, animal-, or plant-based. Materials from natural sources such as collagen derived from animal tissues appear to be advantageous because of their inherent properties of biological recognition. However, some concerns are associated with the use of biologically derived materials, particularly with respect to the complexities associated with purification, sustainability of production, immunogenicity, and pathogen transmission. In addition, medical applications of natural bioabsorbable polymers are limited because their mechanical strength is reduced when they are hydrated. Most natural polymers are soluble in aqueous media unless cross-links are introduced.

Proteins

The major source of naturally derived proteins is bovine or porcine connective tissue from the pericardia, blood vessels, heart valves, and intestine [1]. They are typically tough and pliable, in contrast to synthetic, hydrophilic polymers. The pronounced toughness and pliability of naturally derived materials despite their high water content, e.g., 70–80%, are due largely to the organized orientation of collagen fibrils and the presence of cross-linked elastin in the connective tissue. Drawbacks of the naturally derived materials include possible risks of transmission of viruses and prions such as those involved in bovine spongiform encephalitis (BSE), immunogenicity of persisting cells, potential pathogenicity of their remnants, and biopolymers themselves. However, many natural polymers are able to promote cell attachment due to the presence of cell adhesive sequences.

Collagen

Collagen is the most abundant protein within the ECM of connective tissues such as skin, bone, cartilage, and tendon. Collagen contains large amounts of glycine (nearly 1 in 3 residues, arranged every third residue), proline, and 4-hydroxyproline residues. A typical structure is -Ala-Gly-Pro-Arg-Gly-Glu-4Hyp-Gly-Pro-. At least 20 distinct types of collagen have been identified, and the primary structural collagen in mammalian tissues is type I. As demonstrated in Fig. 3.4, collagen is composed of triple helices of protein molecules that wrap around one another to form a three-stranded rope structure. The strands are held together by both hydrogen and covalent bonds, while collagen strands can self-aggregate to form stable fibers. This highly

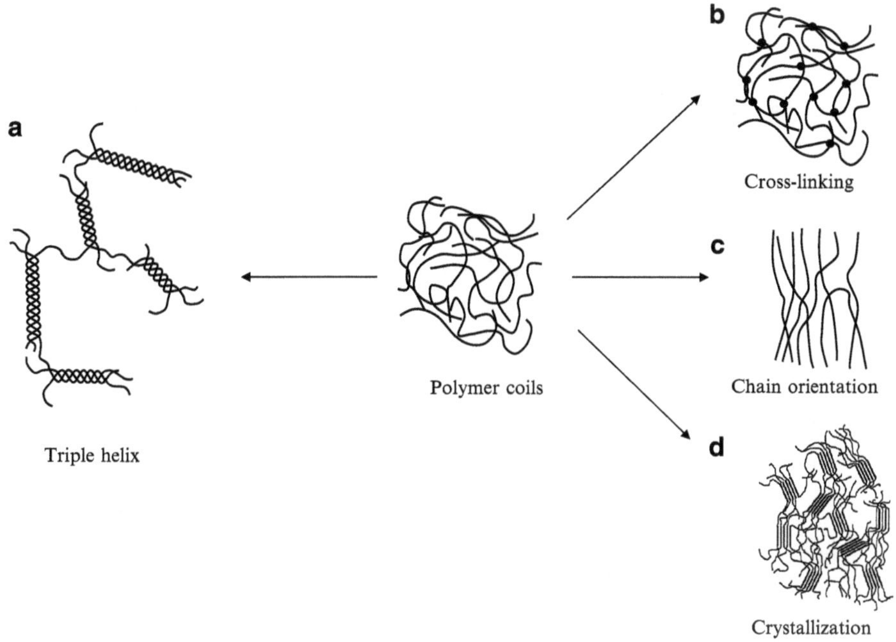

Fig. 3.4 Methods to enhance the mechanical strength of polymers and make them insoluble

ordered structure is one reason why collagen matrix exhibits high mechanical strength. Effective methods of enhancing the mechanical properties of polymers are presented in Fig. 3.4. Collagen is naturally degraded by metalloproteases, specifically collagenase, and serine proteases, permitting its degradation to be locally controlled by cells present in tissues.

Most of the telopeptide portions present at the ends of collagen molecules with antigenic epitopes are removed during the extraction processes. The low mechanical stiffness and rapid biodegradation of extracted but untreated collagen are crucial problems that limit the medical use of this excellent biomaterial. Since cross-linking is an effective method of improving the biodegradation rate and the mechanical properties of polymers, cross-linking treatments have become one of the most important issues in use of collagen. Currently, two kinds of cross-linking methods are applied to collagen, chemical and physical. The physical methods that do not introduce any potential cytotoxic chemical residues include photooxidation, UV irradiation, and dehydrothermal treatment (DHT). Chemical methods are generally applied when higher degrees of cross-linking than those provided by the physical methods are needed. The reagents used in the chemical cross-linking include glutaraldehyde (GA) and water-soluble carbodiimides (WSC), such as 1-ethyl-3-(3-dimethylaminopropyl)-carbodiimide (EDAC). GA, a bifunctional reagent bridging amino groups between two adjacent polypeptide chains through Schiff base formation, is the predominant choice for collagen cross-linking because of its good water solubility, high cross-linking efficiency, and low cost, although GA is a potentially cytotoxic aldehyde. Carbodiimides (CDIs) have been widely used for activation of carboxyl groups of natural polymers under the acidic conditions required for protonation of the carbodiimide nitrogens, leading to nucleophilic attack of the carboxylate anion at the central carbon to form an initial *O*-acylisourea. Although cross-linking by UV and DHT does not introduce toxic moieties into the material, both treatments result in some denaturation of collagen and do not yield enough cross-linking to meet the biomedical demand for collagen. Chemical treatments are therefore preferentially applied with use of traditional GA, WSC, or other methods. Such chemical cross-linking has been shown to markedly reduce biodegradation of collagen.

Besides collagen, elastin plays a major role in determining the mechanical performance of native tissues. Elastin fibers can extend 50–70% under physiological loads, and depending on the location in the blood vessel, elastin content can range from 33 to 200% of that of collagen. In native tissues, elastin exists in stable fibers that resist both hydrolysis and enzymatic digestion.

Gelatin

Gelatin is obtained by extraction of collagen in animal tissues including bone, skin, and tendon, resulting in collagen denaturation. Alkaline extraction of collagen, which converts asparagine and glutamine residues to their respective acids, produces acidic gelatins with isoelectric points below 7, while extraction with diluted acid or enzymes yields basic gelatins with isoelectric points higher than 7. Gelatin is a heterogeneous mixture of single and multistranded polypeptides, each with extended left-handed proline helix conformations containing between 300 and 4,000 amino acids. The triple helix of type I collagen extracted from skin and bone is composed of two a1 (I) and one a2 (I) chains, each with a molecular weight (MW) of ~95 kDa, width of ~1.5 nm, and length of ~0.3 mm. Gelatin consists of mixtures of these strands together with their oligomers as well as degraded (and other types of) polypeptides (see Fig. 3.4). The aqueous solutions undergo coil-helix transition followed by aggregation of the helices under formation of collagen-like right-handed triple-helical proline/hydroxyproline-rich junction zones. Higher levels of these pyrrolidines result in stronger gels. Each of the three strands in the triple helix requires 25 residues to complete one turn; typically there is between one and two turns per junction zone. Gelatin films containing greater triple-helix content swell less in water and consequently have much higher mechanical strength.

Gelatin has been used for a wide range of medical applications because it is readily processed and exhibits good bioabsorbability. Aqueous solutions of gelatin form gels through hydrogen bonding below room temperature and recover sol state upon increase in temperature, which eliminates hydrogen bonding. This reversible sol–gel transition facilitates the molding of gelatin into definite shapes such as blocks and microspheres, while chemical cross-linking is required when dissolution in aqueous media at body temperature should be avoided.

GA has most frequently been used for chemical cross-linking of gelatin to link lysine to lysine, as is true of other proteins as well. Figure 3.5 shows gelation time

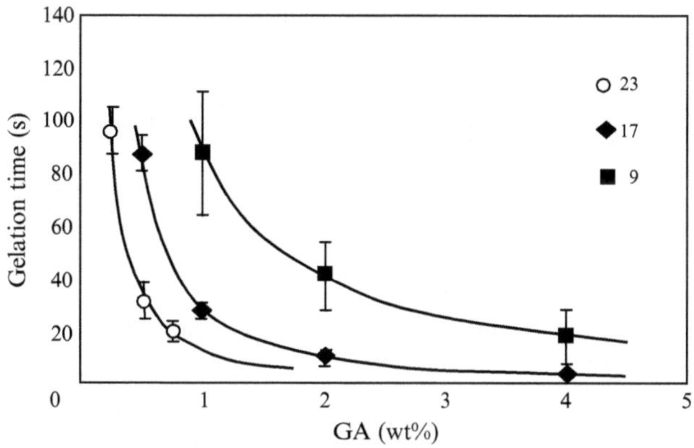

Fig. 3.5 Dependence of gelation time on glutaraldehyde (GA) concentration for gelatin solutions of 9, 17, and 23 wt.% (gelatin was cross-linked at 45°C)

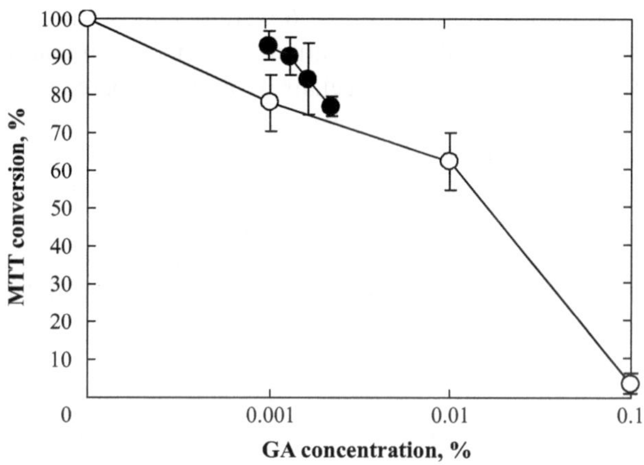

Fig. 3.6 Cytotoxicity of GA (*open circle*), and the gel extract (*closed circle*) evaluated with the MTT assay for human fibroblast

as a function of GA concentration when 9–23 wt.% gelatin solutions are cross-linked at 45°C by mixing the gelatin solutions with GA at low concentration. The concentrations of gelatin and GA in Fig. 3.5 are expressed as those of each prior to mixing of the solution. Clearly, gelatin solutions of higher concentration require GA solution of lower concentration at the same gelation time. One major concern of use of GA as a cross-linking agent is the toxicity of this aldehyde. One method widely used to assess toxicity is cell culture with medium containing the potentially toxic compound. The result of cell culture with assessment of GA with the MTT method is shown in Fig. 3.6. MTT is a yellow compound with the chemical formula of 3-(4,5-dimethylthiazol-2-yl)-2,5-diphenyltetrazolium bromide. This is reduced to purple formazan in the mitochondria of living cells. The absorbance of this colored solution can be quantified by measurement at a certain wavelength with a spectrophotometer. This reduction takes place only when mitochondrial reductases are

Table 3.4 MTT conversion of hydrogel extracts and their GA concentrations

	GA concentration (wt.%)			
Hydrogel	Initial solution	After hydrogel formation	The extract solution[a]	MTT conversion (%)
Gelatin[b]	1	0.1	0.0009	93.0
	2	0.2	0.0013	90.2
	3	0.4	0.0016	84.0
	5	0.6	0.0022	76.9
Serum albumin[c]	10	2.0	0.0257	20.9

[a]Two minutes after the polymerization, 1 g of hydrogel was immersed into 5 ml of buffer solution for 24 h at 37°C. Aldehyde concentration of the extract solution was measured by MBTH method using GA solutions as standards

[b]26 wt.% gelatin and GA solutions were mixed at a volume ratio of 5–0.67, respectively

[c]45 w/v% serum albumin and GA solutions were mixed at a volume ratio of 4–1, respectively

active and conversion can therefore be directly related to the number of viable (living) cells. The results shown in Fig. 3.6 were obtained with human fibroblast culture for 24 h at different GA concentrations. It can be seen that 0.001 wt.% GA results in 77.8% viability. When 1 g of gelatin hydrogel prepared from a mixed solution of 26 wt.% gelatin and 1 wt.% GA was extracted with 5 ml of buffer solution at 37°C for 24 h and the extracted solution was subjected to MTT assay, MTT conversion was found to be 93.0% ($n=3$). As shown in Table 3.4, the GA concentration of extracted solution was 0.0009 wt.%. Increasing the initial GA concentration also increased the aldehyde concentration of the extracted solution. In contrast, MTT conversion was 20.9% for the serum albumin hydrogel prepared from 45 w/v% albumin and 10 wt.% GA solutions.

Fibrinogen and Cross-Linked Polymer Derived from Fibrinogen (Fibrin)

Fibrin is a product of partial hydrolysis by thrombin of fibrinogen. Upon cross-linking, it is converted to a gel. Human fibrin glues have been approved and are available in most major geographical regions of the world. Fibrin is applied to patients as a liquid and solidifies shortly thereafter in situ. Furthermore, fibrin gel readily infiltrates cells since most migrating cells locally activate the fibrinolytic cascade.

Polysaccharides

In addition to proteinaceous materials, naturally occurring polysaccharides and their derivatives have been employed for biomaterial fabrication. Among them are hyaluronic acid, chitin, chitosan, alginates, and agarose. The chemical structures of the repeating units of these polysaccharides are shown in Fig. 3.7. It should be noted that water-soluble alginates and agarose are not biodegradable in the human body.

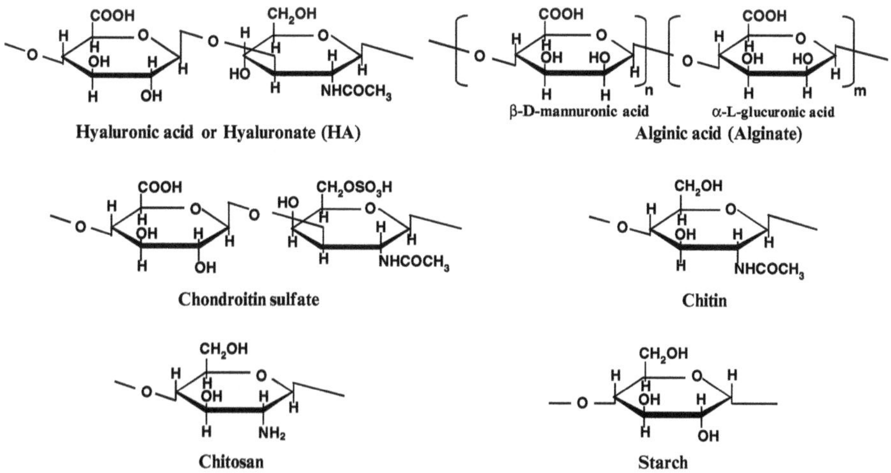

Fig. 3.7 Chemical structures of the polysaccharides used in medical applications

Hyaluronic Acid (Hyaluronate)

Industrially, hyaluronic acid (HA) is obtained from animal tissues such as umbilical cord, cock's comb, vitreous body, and synovial fluid. Biotechnological methods also yield HA on a large scale. HA is the only nonsulfated glucosaminoglycan (GAG) that is present in all connective tissues as a major constituent of ECM and plays pivotal roles in wound healing. As shown in Fig. 3.7, this linear, nonadhesive polysaccharide consists of repeating disaccharide units (β-1,4-D-glucuronic acid and β-1,3-*N*-acetyl-D-glucosamine) with a weight average molecular weight (Mw) of up to 10,000 kDa. This anionic polymer is also a major constituent of the vitreous (0.1–0.4 mg/g), synovial joint fluid (3–4 mg/mL), and hyaline cartilage, where it reaches approximately 1 mg/g wet weight. Clearance of HA from the systemic circulation results in a half-life of 2.5–5.5 min in plasma. In solution, HA assumes a stiffened helical configuration due to hydrogen bonding, and the ensuing coil structure traps approximately 1,000-fold weight of water. The highly viscous aqueous solutions thus formed give HA unique physicochemical and biological properties that make possible preservation of tissue hydration, enable regulation of tissue permeability through steric exclusion, and permit joint lubrication. In the ECM of connective tissues, HA plays important roles in maintaining tissue morphologic organization, preserving extracellular space, and transporting ions, solutes, and nutrients. Along with ECM proteins, HA binds to specific cell surface receptors, and the resulting activation of intracellular signaling events leads to ECM stabilization, regulates cell adhesion and mobility, and promotes cell proliferation and differentiation. HA signaling also occurs during morphogenesis and embryonic development, during modulation of inflammation, and in the stimulation of wound healing. Corresponding to these multiple functions, HA is a strong inducer of angiogenesis, although its biological activity in tissues has been shown to depend on

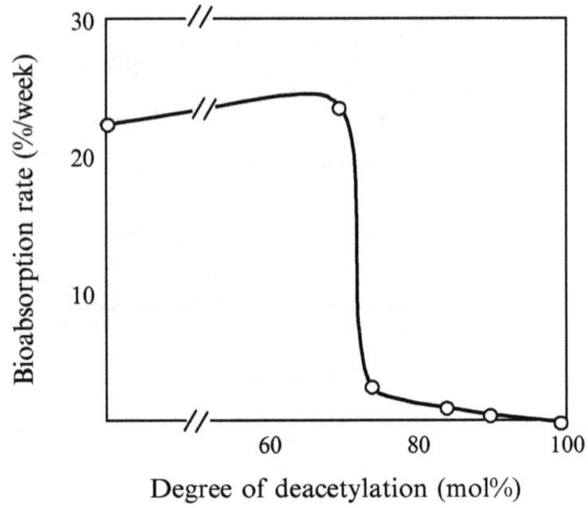

Fig. 3.8 Dependence of the initial bioabsorption rate of films obtained from chitin and its deacetylated derivatives on the degree of deacetylation. Reproduced from [2] with permission from Elsevier

molecular size. High-molecular-weight native-HA (n-HA) has been shown to inhibit angiogenesis, whereas degradation products of low-molecular-weight HA stimulate endothelial cell proliferation and migration. HA is naturally hydrolyzed by hyaluronidase, allowing cells in the body to regulate and limit clearance of it.

Due to its unique physicochemical properties, unmodified HA has been widely used in the fields of viscosurgery, viscosupplementation, and wound healing. However, the poor mechanical properties of this water-soluble polymer and its rapid degradation in vivo have precluded its wide clinical application as a biomaterial. Therefore, in an attempt to obtain materials that are more mechanically and chemically robust, a variety of covalent cross-linking via hydroxyl or carboxyl groups, esterification, and annealing strategies have been explored to produce insoluble HA hydrogels. However, the cross-linking agents used are often cytotoxic small molecules, and the resulting hydrogels must be extracted or washed extensively to remove traces of unreacted reagents and by-products.

Chitosan and Chitin

Chitosan is a linear polysaccharide of (1–4)-linked D-glucosamine and *N*-acetyl-D-glucosamine residues derived from chitin, which is found in arthropod exoskeletons (Fig. 3.7). The degree of N-deacetylation of chitin usually varies from 50 to 90% and determines crystallinity, which is the greatest for 0 and 100% N-deacetylation. Chitosan is soluble in dilute acids, which protonate the free amino groups. Once dissolved, chitosan can be gelled by increasing the pH or extruding the solution into a nonsolvent. Chitosan derivatives and blends are also gelled via GA cross-linking, UV irradiation, and thermal variation. Chitosan is degraded by lysozyme, and its kinetics of degradation are inversely related to the degree of crystallinity. Figure 3.8 shows the dependence of bioabsorption of chitin on the extent of hydrolysis when partially hydrolyzed chitin (or partially acetylated chitosan) is subcutaneously implanted in rat [2]. It is seen that 100% chitin and 100% chitosan do not undergo in vivo bioabsorption, at least in rats.

Synthetic Polymers

Before the prion shock, naturally derived materials attracted much attention because of their natural origin, which seemed to guarantee biocompatibility. However, reports on Creutzfeldt-Jakob disease due to implanted sheets made from human dried dura mater diverted the focus of biomaterial scientists to nonbiological materials such as synthetic polymers. In addition, synthetic biomaterials have received a great deal of attention in the medical field due to their precisely controlled processes of manufacturing and uniform and reproducible properties.

Biodegradation can be conferred to polymers by synthesis with hydrolytically unstable linkages, as shown in Table 3.3, in the backbone. This is commonly achieved with the use of chemical functional groups such as ester anhydrides, orthoesters, and amides. Most biodegradable polymers are synthesized by ring-opening polymerization. Among the most common biodegradable polymers produced by ring-opening polymerization are poly(α-hydroxyacid)s, including PGA, PLA, and poly-*p*-dioxanone (PDS). For simplicity, synthetic biodegradable polymers are here classified into two groups, aliphatic α-polyesters and others.

Poly(α-Hydroxyacid)s [Aliphatic α-Polyesters or Poly(α-Hydroxyester)s]

The majority of biodegradable, synthetic polymers that are currently available are poly(α-hydroxyacid)s that have repeating units of –O–R–CO–(R; aliphatic) in the main chain. This is mainly because most of them have the potential to produce biomaterials with good mechanical properties and have been approved by the US FDA for a variety of clinical applications as bioabsorbable medical devices with biosafe by-products of degradation. Poly(α-hydroxyester)s are thermoplastic polymers with hydrolytically labile aliphatic ester linkages in their backbone. Although all polyesters are theoretically degradable, only aliphatic polyesters with reasonably short aliphatic chains between ester bonds can exhibit degradation over the time frame required for biomaterials. Aromatic polyesters with phenyl groups in the main chain (which are materials used for vascular grafts) do not undergo any appreciable degradation under physiological conditions. The monomers used for synthesis of poly(α-hydroxyacid)s include glycolic acid (or glycolide) and L- and DL-lactic acids (or -lactides) (LLA or DLLA) with a hydroxyl group on the α carbon. These monomers can yield not only homopolymers but also copolymers when polymerized with other monomers such as ε-caprolactone, *p*-dioxanone, and trimethylene carbonate. The chemical structures of α-hydroxyacid monomers are shown in Fig. 3.9. PGA was discovered in the 1960s and first implanted in the 1970s.

Homopolymers

The most widely used bioabsorbable sutures are made from PGA or poly(glycolide-co-lactide) (PGLA) with a glycolide/L-lactide ratio of 90/10. This PGLA with 90% glycolide is commercially available as a multifilament suture and a mesh

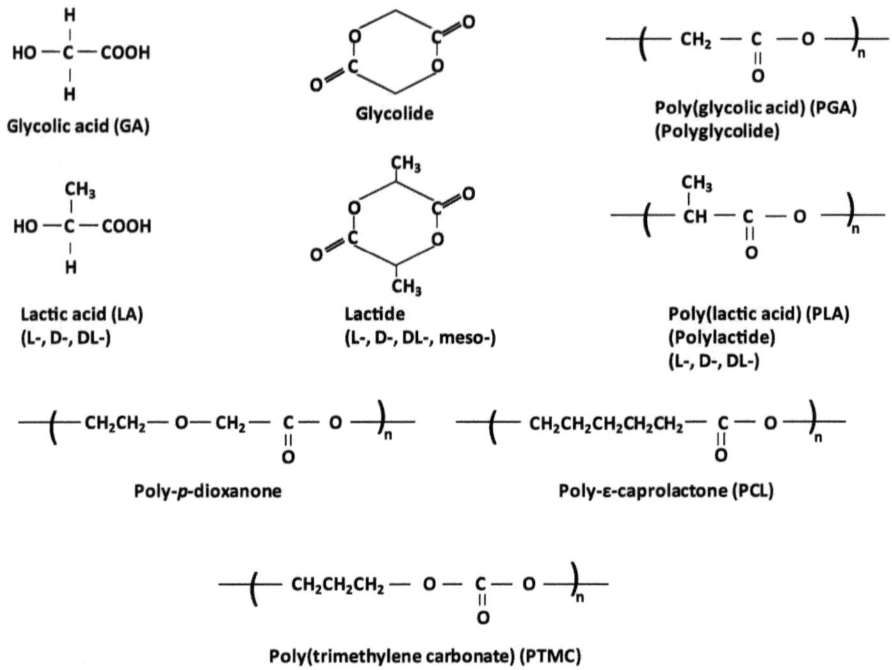

Fig. 3.9 Chemical structures of α-hydroxyacid polymers, copolymers, and their monomers

(Vicryl, Ethicon, USA). PGA is the simplest linear aliphatic polyester. Because its hydrolytic degradation is controllable, PGA and its copolymers with LA, ε-CL, and TMC are widely used as bioabsorbable materials. PGA is a highly crystalline (around 45–55%) polymer with a glass transition temperature between 35 and 40°C and a melting point in the range of 225–230°C. It is soluble only in highly fluorinated solvents like hexafluoroisopropanol and hexafluoroacetone sesquihydrate that can be used to prepare solutions of high-MW polymer for spinning and film preparation. Fibers of PGA exhibit high strength and Young's modulus due to its high crystallinity. Poly(L-lactide) (PLLA) has been clinically used after molding into pins, screws, and miniplates for fixation of fractured bones of patients. PGA and PLLA are crystalline polymers that can provide medical devices with excellent mechanical properties, but PGA undergoes degradation very quickly, while PLLA is degraded very slowly when implanted. Crystalline biodegradable polymers such as PGA and PLLA undergo hydrolytic degradation through nonspecific scission of the ester backbone involving several steps; the first step involves diffusion of water into the amorphous (noncrystalline) regions of the polymer matrix, with cleaving of the ester bonds, with the second step starting after the amorphous regions have been eroded, leaving the crystalline portion of the polymer susceptible to hydrolytic attack.

It takes longer than 1 year, and usually a few years, for homopolymers of L-lactide and ε-caprolactone to be completely bioabsorbed. In addition, trimethylene carbonate homopolymer exhibits very slow degradation in the presence of water.

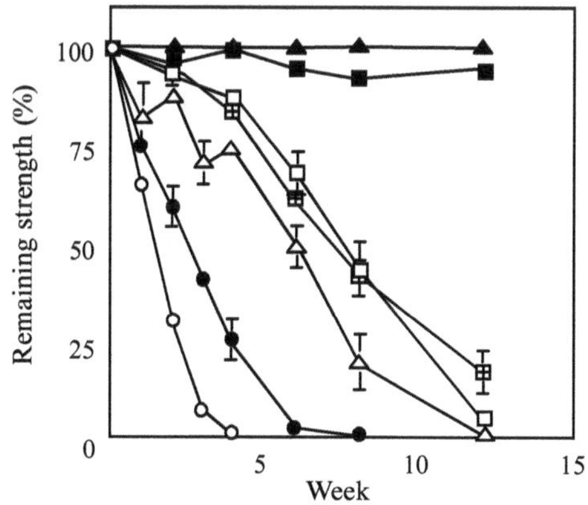

Fig. 3.10 Change in the tensile strength of bioabsorbable fibers on immersion in buffer solution (pH 7.4) at 37°C. *Open circles*, glycolide/ε-caprolactone (75:25); *open squares*, *p*-dioxanone (100); *open triangles*, glycolide/trimethylene carbonate (67.5:22.5); *solid circles*, glycolide/dioxanone/trimethylene carbonate (60:14:26); *crossed squares*, L-lactide/ε-caprolactone (80:20); *solid squares*, caprolactam (100); and *solid triangles*, propylene (100) [3]

Homopolymers of DL-lactide (PDLLA) are bioabsorbed at a slightly higher rate than PLLA, due to the absence of crystalline regions. When it is not necessary to distinguish between PLLA and PDLLA, the term PLA is used to represent both PLLA and PDLLA.

Copolymers

Copolymerization of glycolide and L-lactide with other comonomers is used to modify the properties of PGA and PLLA and to control their degradation behavior to make it suitable for specific applications in the medical field. Figure 3.10 shows decrease in tensile strength in phosphate buffer solution versus time of hydrolysis for various aliphatic polyesters in fiber form [3]. Copolymerization of monomers A and B offers great potential for modification of polymers A and B by controlling the physical and biological properties of biodegradable polymers, such as rate of degradation, hydrophilicity, mechanical properties, and in vivo shrinkage. Since clinicians are apt to confuse the definition of copolymer with that of polymer blend, the difference between the copolymers and blends of them is briefly described below.

Assume that homopolymer A is bioabsorbed while homopolymer B exhibits no or only insignificant bioabsorption in the body. Simple blending of homopolymers A and B does not change the degradation kinetics of each homopolymer, but copolymerization of monomer B with monomer A converts homopolymer B to a biodegradable copolymer. This scheme is illustrated in Fig. 3.11 for the equimolar copolymerization of monomers A and B. The continuous sequence of each monomer in the copolymer chain is governed by the monomer reactivity and polymerization

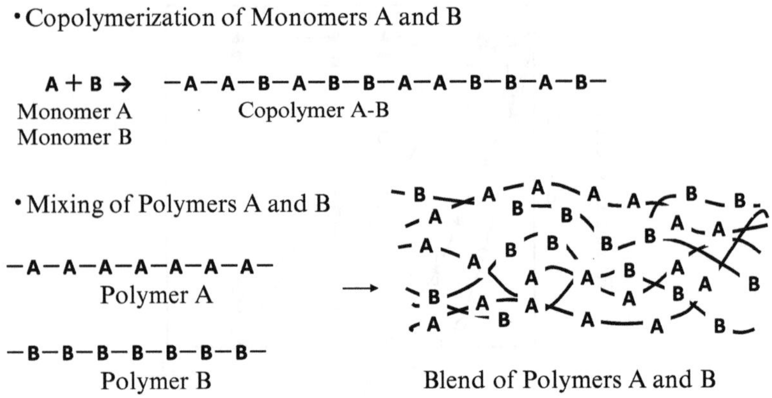

Fig. 3.11 Difference between copolymer A/B and a blend of polymers A and B

conditions such as initiator concentration and temperature. If the continuous sequence of monomer B in the copolymer chain is shorter than a critical length below which oligomers B are soluble or dispersible in aqueous media, the copolymer A–B becomes biodegradable due to the degradation of the monomer A unit sequence. A typical example is lactide-ε-caprolactone copolymers, which are bioabsorbed in the body at rates higher than those of both the lactide homopolymer and the ε-caprolactone (CL) homopolymer, which is virtually nonbioabsorbable. Furthermore, this copolymerization converts the brittle lactide homopolymer into a much more rubber-like, tough polymer. Figure 3.12 shows how copolymerization of LLA with CL yields copolymers with low Young's moduli [4] and high bioabsorption rates. It is possible that DLLA copolymerization with CL produces copolymers with properties different from those of LLA-CL copolymers since the long sequences predominant in the LLA chains will result in formation of small crystals via association among themselves, in marked contrast to DLLA sequences, which do not have any potential to crystallize.

Copolymerization of DLLA or CL with 1,3-trimethylene carbonate (TMC) has also been attempted. High-molecular-weight TMC homopolymer is an amorphous elastomer that exhibits good mechanical performance, combines high flexibility with high tensile strength, and degrades very slowly at pH 7.4 and 37°C. High-molecular-weight copolymers of TMC and DLLA with 20–50 mol% of TMC are amorphous, relatively strong elastomers that can maintain mechanical properties up to 3 months at in vitro rates of degradation and are bioabsorbed in less than 1 year. In contrast, copolymers of TMC and CL are degraded more slowly than TMC-DLLA copolymers. TMC-CL copolymers with high CL content are semicrystalline, very flexible, and tough and can therefore maintain mechanical properties for more than 1 year when incubated in buffer solution at pH 7.4 and 37°C.

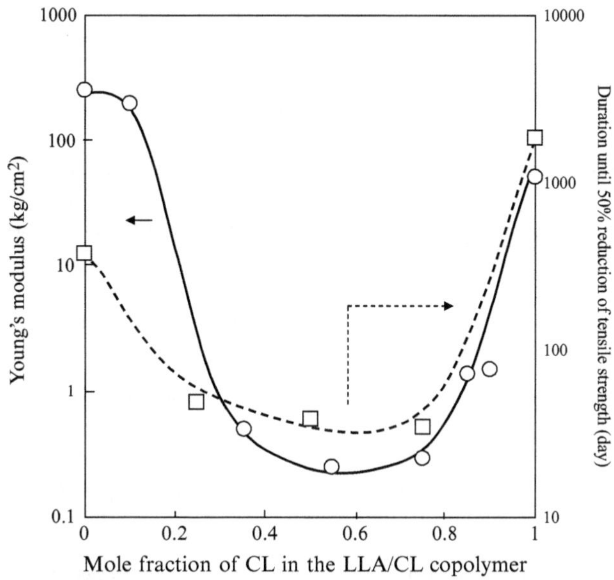

Fig. 3.12 The Young's modulus and hydrolysis time (until 50% reduction of tensile strength) for the L-lactide/ε-caprolactone (LLA/CL) copolymer plotted against the mole fraction of the CL comonomer in the copolymer

Others

In addition to aliphatic polyesters, a number of synthetic, biodegradable polymers have been studied for medical application. The chemical structures of these polymers are shown in Fig. 3.13. Only polyurethane will be presented below since this polymer has been studied by several research groups. Since polyesters are limited with respect to ability to withstand high strain and elastic capabilities, interest has been paid to the development of biodegradable elastomers with high elastic capability and mechanical strength. Segmented polyurethane elastomers have been used for elastic devices such as indwelling catheters, intra-aortic balloons, and left ventricular assist devices, but they are not biodegradable. Segmented polyurethanes are basically synthesized from a polymer with a terminal hydroxyl group at both chain ends (HO–P–OH) and an excess of diisocyanate (OCN–R–NCO). Upon mixing them under an excess of diisocyanate, a prepolymer with terminal isocyanate groups is produced in association with the formation of urea bonds between –OH and –NCO. Addition of a diamine chain extender to this prepolymer increases the chain length to produce segmented polyurethane. Association with the P portion yields a soft segment while the R portion forms a hard segment. If the P portion comprises hydrolysable units, the resultant polyurethane exhibits biodegradation.

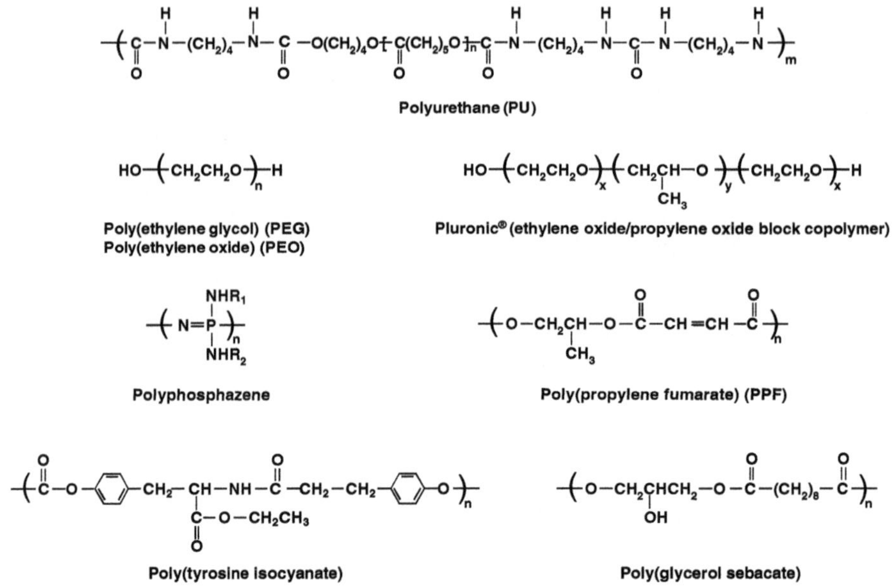

Fig. 3.13 Chemical structures of synthetic, bioabsorbable polymers, excluding those of α-hydroxyacid polymers

Water-Soluble Polymers

There is one group of polymers that is not biodegradable and is instead bioabsorbable in the body. The polymers in this group are water-soluble, unless cross-linked or blended. Typical examples include carboxymethyl cellulose (CMC), a natural polymer, and poly(ethylene glycol) (PEG), a synthetic polymer.

CMC

CMC is obtained by alkali-catalyzed carboxymethylation of cellulose at the C6 site with chloroacetic acid. This reaction changes water-insoluble cellulose into a water-soluble polymer, depending on the substitution reaction involved. This modified cellulose is used in industry primarily because it has high viscosity, is nontoxic, and is nonallergenic. CMC is medically used as a lubricant in nonvolatile eye drops (artificial tears). CMC becomes water-insoluble when esterified between the COOH group and the OH group of CMC or other polymers possessing COOH and OH groups with the activating agent 1-(3-dimethylaminopropyl)-3-ethylcarbodiimide hydrochloride (EDC).

Alginic Acid (Alginate)

Alginates are linear polysaccharides derived primarily from brown seaweed and bacteria. They are block copolymers composed of regions of sequential (1–4)-linked β-D-mannuronic acid monomers (M-blocks), regions of α-L-guluronic acid (G-blocks), and regions of interspersed M and G units (Fig. 3.7). The lengths of M- and G-blocks and their sequential distribution along the polymer chain vary depending on the source of alginates. These biopolymers undergo reversible gelation in aqueous solution under mild conditions through interaction with divalent cations including Ca^{2+}, Ba^{2+}, and Sr^{2+} which can cooperatively bind between the G-blocks of adjacent alginate chains, creating ionic interchain bridges. This highly cooperative binding requires more than 20 G-monomers.

Gels can also be formed by covalently cross-linking alginate with adipic hydrazide and PEG using standard carbodiimide chemistry. Ionically cross-linked alginate hydrogels do not undergo specific degradation and undergo slow, uncontrolled dissolution instead. Alginate-Ca^{2+} is lost through ion exchange of calcium with sodium, followed by dissolution of individual chains, which results in loss of mechanical stiffness over time. Alginates are easily processed into any desired shape with the use of divalent cations. One possible disadvantage of use of alginates is their low and uncontrollable in vivo bioabsorption rate, which is principally the result of the sensitivity of these gels to calcium chelating compounds (e.g., phosphate, citrate, and lactate). Several in vivo studies have demonstrated large variations in the rate of degradation of calcium-cross-linked sodium alginate.

PEG

Cross-linking of water-soluble polymer chains produces water-insoluble 3-D networks. The product obtained is termed hydrogel. Several distinctive features, including tissue-like viscoelasticity, diffusive transport, and interstitial flow characteristics, make synthetic hydrogels excellent physicochemical mimetics of natural ECMs and candidates for soft-tissue scaffolds. Indeed, hydrogels with high water content have the potential to efficiently encapsulate cells and to allow nutrient and waste transport. Soft tissues in the human body are similar to hydrogels in having high water content ranging between 70 and 90%. The structural integrity of hydrogels depends on cross-links formed between polymer chains via physical, ionic, or covalent interaction. Synthetic hydrogels can be processed under relatively mild conditions, have mechanical and structural properties similar to ECM, and can be delivered into the body in minimally invasive fashion, particularly by injection. A variety of synthetic materials including PEG, poly(vinyl alcohol)(PVA), and poly(acrylic acid) can be used to form hydrogels. Figure 3.14 shows their chemical structures. These polymers are not biodegradable and are instead water-soluble unless cross-links are introduced. When the cross-links that have been introduced are broken, the resulting

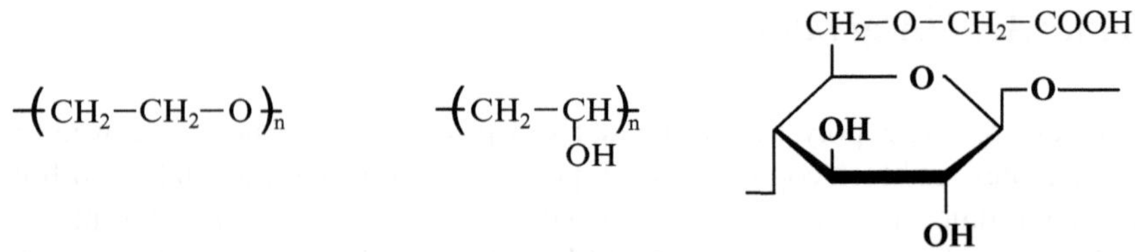

Fig. 3.14 Chemical structures of water-soluble polymers studied for medical application

polymer chains again become water-soluble and bioabsorbable in the body and are excreted via the kidney or bile duct, depending on molecular size. Naturally derived polymers including collagen, gelatin, fibrinogen, albumin, polypeptides, HA, agarose, alginate, and chitosan can also be used to obtain hydrogels.

Among synthetic polymers, PEG, or poly(ethylene oxide)(PEO) with the same basic chemical structure as PEG, is most commonly used for hydrogel fabrication in the biomedical area and is currently approved by the FDA for several medical applications. Each end of star-shaped PEG chains can be modified with either acrylates or methacrylates to facilitate photo-cross-linking. Hydrogels form when the modified PEG is mixed with an appropriate photoinitiator and cross-linked via UV exposure, as shown in Fig. 3.3. This method of photoinduction to produce hydrogels offers diverse, minimally invasive applications via arthroscopy/endoscopy or subcutaneous injection for tissue replacement or augmentation. Thermally reversible hydrogels have also been created from block copolymers of PEG and PLLA.

References

1. Crapo PM, Gilbert TW, Badylak SF (2011) An overview of tissue and whole organ decellularization processes. Biomaterials 32:3233–3243
2. Tomihata K, Ikada Y (1997) In vitro and in vivo degradation of films of chitin and its deacetylated derivatives. Biomaterials 18:567–575
3. Tomihata K, Suzuki M, Ikada Y (2001) The pH dependence of monofilament sutures on hydrolytic degradation. J Biomed Mater Res 58:511–518
4. Hyon S-H, Nakamura T, Cha W-I, Ikada Y (1996) Synthesis and properties of elastomeric lactide-caprolactone copolymers. J Japanese Soc Biomater 14:216–223

Chapter 4
Sealants (Adhesives) to Prevent Bleeding

Abstract In general, surgical sealants (adhesives) are liquids and form gels when applied. Currently, several products are clinically available, including naturally derived fibrin sealants, semisynthetic products such as gelatin and albumin-based glues, and synthetic products including cyanoacrylates and poly(ethylene glycol). These products, however, have both advantages and disadvantages. While fibrin sealants have been successfully used in many surgical fields, they have several disadvantages including low adhesive strength. Other products, such as cyanoacrylates and protein-based products, have high adhesion strength but have high toxicity, and therefore, their applications are limited. Studies are still ongoing in order to develop sealants with higher adhesion strength and safer properties compared to the sealants that are currently available. This chapter reviews the chemistry and applications of clinically available sealants and recent developments in this field.

Introduction

Surgical sealants (adhesives) have been widely used in surgery to prevent air leaks (from holes in diseased soft tissues such as lung), liquid leaks (including hemostasis, as for oozing), and as adhesives (to bond two separate tissues, but very rarely). They have many advantages over traditional techniques (i.e., sutures and staples) such as reducing both operative time and physical load on patients. Sutures have been most extensively used for wound closure, though they have shortcomings such as the highly skilled procedures involved, the long time required for wound closure, and the postoperative removal of nonbioabsorbable sutures [1]. Alternatively, electrocauterization, infrared coagulation, compression, and tamponade can be used for wound closure, especially for prevention of bleeding. However, very small blood vessels, fragile tissues, and small spaces make application with these devices difficult. Sealants are useful in these cases.

S. Suzuki and Y. Ikada, *Biomaterials for Surgical Operation*,
DOI 10.1007/978-1-61779-570-1_4,

They are often liquids and cure when applied to form a firm gel. Ideally, they should be rapidly curable even in the presence of water, as pliable as soft tissues when cured, and bioabsorbable in the body. Three main types are available for these materials: naturally derived (such as fibrin glue), semisynthetic (such as naturally derived polymer—aldehyde-based), and synthetic (such as cyanoacrylates). In this chapter, commercial products of these materials are discussed, followed by recent developments in surgical sealants and adhesives.

Requirements for Surgical Sealants (Adhesives)

Requirements for these surgical sealants (adhesives) are listed in Table 4.1 [1]. As they are applied to the human body, they must be safe, nontoxic, and free from risk of infectious transmission. They should be bioabsorbable and not persist long as a foreign body. The sealants (i.e., solutions that are applied and form a gel) as well as their degradation products must be safe. In addition, they must not hinder the natural healing process.

Most surgical sealants are gelled in situ, are applicable on rough surfaces, and act as tissue anchors. Appropriate gelling time and mechanical properties of the formed gel are important. They should bond rapidly to the surrounding living tissues in the presence of water or aqueous fluids such as blood and maintain adhesive strength. Mechanical properties of the cured materials are important, as they may induce undesirable tissue reactions if the cured glue is harder than the soft tissue. Good handling, such as ease of use and gelation speed, is also important. Several commercially available products are listed in Table 4.2.

Naturally Derived Sealants

Fibrin Glue

Fibrin glue is based on the natural clotting process. It became commercially available in the 1970s in Europe and has been widely used in Europe, Canada, and Japan. In the United States, fibrin glue was banned until 1998, because the USFDA was

Table 4.1 Requirements for surgical sealants [1]

1.	In situ curable from the liquid state through polymerization, chemical cross-linking, or solvent evaporation
2.	Rapidly curable under wet physiological conditions
3.	Strongly bondable to tissue
4.	Nontoxic agents and degradation products
5.	As tough and pliable as natural tissues
6.	Bioabsorbable

Table 4.2 Commercially available sealants, adhesives, and hemostatic agents

Structure	Name	Main composition	Company
Sealants	Tisseel, Artiss	Fibrinogen and thrombin from human pooled plasma	Baxter
	Evicel		JJ
	Beriplast		Aventis Behring
	DuraSeal, DuraSeal Xact	PEG-NHS and trilysine	Confluent surgical
	CoSeal	PEG-NHS and PEG-SH	Baxter
	FocalSeal (AdvaSeal)	PEG	Genzyme
Adhesives	GRF® glue	Gelatin-resorcinol-formaldehyde-glutaraldehyde	Cardial
	BioGlue	Bovine albumin and glutaraldehyde	CryoLife
	Dermabond	Octyl-2-cyanoacrylate	Colure Medical
	Histoacryl	n-Butyl-2-cyanoacrylate	B Braun
	Indermil		Covidien
Adhesive film	TissuePatch3	(VA)-(AAc)-(AAc-NHS) and PGLA	Tissuemed
Hemostatic agents (stand-alone)	Gelfoam	Gelatin	Pfizer
	Surgicel	Oxidized cellulose	Ethicon
	Abiten	Collagen	Bard
	HemCon, ChitoFlex	Chitosan	HemCon Medical
(combination)	TachoComb	Fibrinogen, thrombin, and collagen	Nycomed Arzneimittel
	FloSeal	Gelatin and thrombin	Baxter

concerned about the risk of viral disease transmission in products including components of the pooled-plasma tissue adhesive product. It is composed of human-derived fibrinogen, thrombin, factor XIII, and bovine aprotinin. Aprotinin is a basic pancreatic trypsin inhibitor extracted from bovine lung tissues. It is added to commercial fibrin glues to yield delayed fibrinolysis of clots. However, aprotinin also exhibited concomitant slower regeneration of granulation tissue and was thus found to be ineffective for wound healing. The fibrin clot formation obtained with solutions of fibrinogen (molecular weight 360 kDa) and thrombin involves enzymatic and polymerization reactions, as shown in Fig. 4.1.

First, thrombin catalyzes the partial hydrolysis of fibrinogen chains to form fibrin monomers. These monomers then react with one another in the presence of calcium ions to form polymers that eventually form a three-dimensional gel network. Thrombin also activates factor XIII to factor XIIIa in the presence of calcium ions. Finally, factor XIIIa covalently cross-links α and γ chains of different fibrin monomers in the matrix. The fibrin clot is degraded by physiologic fibrinolysis, which is initiated by a complex cascade of enzymes over a 2-week period [2].

Commercial fibrin glue comes in two component systems; one contains fibrinogen, factor XIII, fibronectin, and fibrinolysis inhibitor, while the other contains

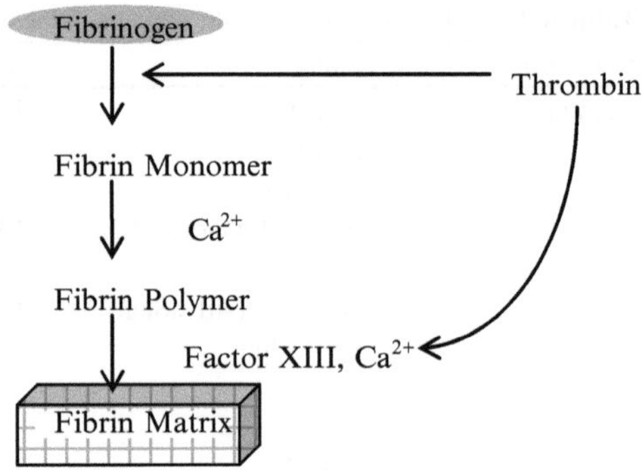

Fig. 4.1 Fibrin clot formation from fibrinogen

Table 4.3 Compositions of fibrin glues

Components	Concentration	Origin
Fibrinogen (clottable)	80–120 g/l	Human plasma
Factor XIII (activity)	10–30 IU/ml	Human plasma or human placenta
Fibronectin (antigen)	5–20 g/l	Human plasma
Thrombin (activity)	300–600 NIH-U/ml	Human plasma or bovine plasma
Aprotinin	3,000 KIU/ml	Bovine lung
Calcium chloride	40–60 mM	Inorganic

thrombin and calcium chloride. The concentrations of these components are shown in Table 4.3. The concentration of fibrinogen determines the strength of the fibrin clot, and the amount of thrombin controls the rate of clot network formation. These are set in a dual-chamber delivery system with or without a spraying device. A fibrin clot is formed within 30 s when the components are applied at the site of injury. The process of preparation requires constant warming and stirring, which takes around 20–30 min.

Fibrin glue has many advantages such as acceptable biocompatibility, biodegradability, injectability, and in situ–producible scaffolds for tissue regeneration. Successful clinical application of fibrin glue has been demonstrated in many surgical fields [3, 4]. There are, however, several disadvantages, such as the risk of viral infection due to human- and animal blood–derived products, low strength, and rapid loss of strength [5]. It also requires long preparation time and operator skill and is quite expensive, at approximately $100–150/ml of fibrinogen. In the case of commercial products, careful donor selection, extensive plasma screening, and effective viral reduction steps during preparation have significantly reduced the possibility of viral transmission. No cases have been reported of serious viral transmission due to commercial fibrin glues [3]. However, the risk of viral infection is not completely zero due to contamination from individuals with acquired immunodeficiency and hepatitis. The recombinant technology has developed new products, which can be considered also for use as fibrin glue. In 2008, recombinant human thrombin,

Table 4.4 Clinical applications of fibrin glue [1]

Cardiovascular surgery	Hemostasis of bleeding from mediastinum and small blood vessels, reinforcement of suture line, sealing of vascular grafts, and anastomoses
Thoracic surgery	Sealing of air leaks from raw lung surface, and pulmonary and bronchial staple lines
Gynecologic surgery	Anastomosis of fallopian tube, adhesion of peritoneum
Plastic surgery	Closure of skin wound, attachment of skin grafts, fixation of tissue flaps
Orthopedic surgery	Fixation of bone fragments, hemostasis during total knee arthroplasty
Urologic surgery	Anastomosis of urinary tube and adhesion of kidney, pelvis renalis, and ureter

Recothrom®, produced by ZymoGenetics (Seattle, WA), was approved by the FDA as a topical hemostatic agent [6], while recombinant human fibrinogen was granted as Orphan Drug Status by the FDA [7].

The clinical applications of fibrin glue and its purposes are summarized in Table 4.4. Cardiovascular surgery is the most common among them, and its major purpose is hemostasis and suture hole sealing. Successful local hemostasis of bleeding after open-heart surgery can reduce blood loss, operative time, and the need for resternotomy in high-risk patients. Fibrin glue is highly effective as a tissue sealant in cases of diffuse, low-pressure bleeding such as coronary vein bleeding, right-ventricle/pulmonary-artery conduits, and right ventricular patches. Fibrin glue is used routinely whenever suturing is impossible, dangerous, or difficult, and for spontaneous hemostasis to save waiting time.

Suture holes, stable lines, anastomoses, fistulas, and raw surfaces are effectively treated with fibrin glue, similar to aortic and coronary suture lines. It is also used for sealing vascular grafts. Perfect hemostasis is important in pediatric cardiac surgery because of the high incidence of coagulopathy after open-heart procedures and the increased risk associated with the use of blood and blood products. It has also been applied to endoscopic procedures for the treatment of peptic ulcers and for prevention of anastomotic leaks during gastric bypass surgery.

Despite the successful application of fibrin glue, many technical efforts have been made to improve its low adhesive strength, especially in situations like sealing of needle holes. Methods of application include a separate dripping method using two separate syringes, a simultaneous dripping method using a syringe applicator, and a spray method using a spray applicator. Although the spray method exhibits the strongest sealing effect due to homogeneous application resulting in complete mixing of the two components [8–10], adhesive strength is still not adequate for control of local bleeding in the clinical setting. The rub-and-spray method consists of rubbing the fibrinogen solution onto the needle holes with a finger and then spraying the two solutions using an application nozzle. This technique has been experimentally and clinically found to be effective in cardiovascular surgery [11, 12].

Fig. 4.2 Photograph of Neoveil

Fibrin Glue in Combination with PGA Sheet (Neoveil)

Although fibrin sealant has been widely used to seal pulmonary air leakage in lung surgery, surgeons have often found that fibrin sealant does not effectively seal air leakages. To overcome this problem, the application of fibrin glue in combination with a poly(glycolic acid) (PGA) sheet (Neoveil, Gunze Corp, Kyoto, Japan) has been developed and shown to have clinical effectiveness [13]. The PGA sheet is a soft nonwoven fabric, which is flexible, easily shaped to reconstruct complex shapes, and highly compatible with fibrin glue (Fig. 4.2). It is degraded by hydrolysis and bioabsorbed in about 15 weeks. The PGA sheet retains fibrin glue that otherwise would run down from the rough site of application. The combination of PGA fabric and fibrin glue has been shown to improve adhesive strength and resistance to air/water leakage. This technique has been expanded to other applications including neurosurgery and gastroenterological and plastic surgeries.

The speed of adhesion and strength of fibrin glue in combination with the PGA sheet have been investigated in vitro [14]. Figure 4.3 shows the burst pressure of the adhesive on a defect (with a diameter of 1 cm) created on rabbit skin. The PGA sheet soaked in fibrinogen solution was placed on the defect, and fibrinogen and thrombin solutions were applied either simultaneously by spray nozzle or separately by syringe. Simultaneous application by spraying created a strong sealant that could bear high pressure (above 220 mm Hg), whereas separate application resulted in weaker sealing. Several application techniques have been compared for application in cases of alveolar air leakage [15].

It is important and essential to successful early postoperative outcome to prevent air leaks after major lung resection for cancer. However, postoperative air leaks occur frequently after closure of the chest wall despite thorough water-seal testing. Despite many investigations of surgical sealants, not enough evidence has been obtained to support their use. Ueda et al. compared patients with postoperative air leakage treated using sutures and fibrin glue ($n=61$) with others treated using PGA mesh and fibrin glue without suturing ($n=61$) [16]. It was found that the duration of chest tube drainage and postoperative hospital stay were significantly shorter in the

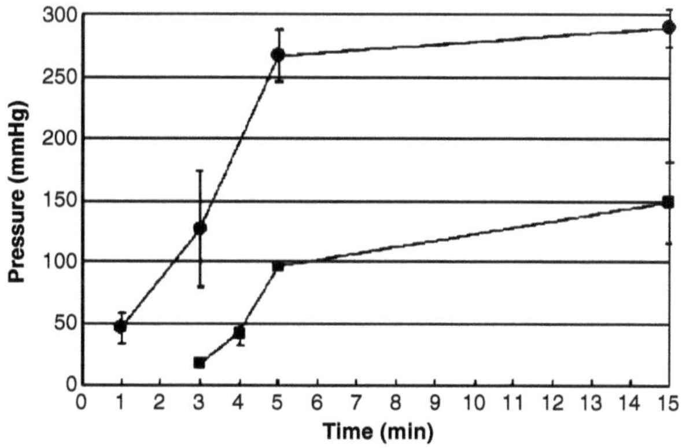

Fig. 4.3 Relationship between reaction time and sustainable pressure. After attaching the poly(glycolic acid) (PGA) sheet to rabbit skin, fibrin glue was applied separately by using a syringe (*squares*) or simultaneously by spraying (*circles*) [14]

Table 4.5 Postoperative complications in the glue-alone (G) and PGA mesh-and-glue (MG) groups [16]

Complications	G group ($n=61$)	MG group ($n=61$)
Pulmonary complications	4(7%)	0(0%)*
Pneumonia	2	0
Respiratory failure	1	0
Lobar atelectasis	1	0
Cardiac complications	2(3%)	2(3%)
Atrial fibrillation	2	1
Pulmonary embolism	0	1
Total	6(10%)	2(3%)

*$p=0.042$ versus the glue-alone group

PGA-glue group than in the glue-alone group and that the incidence of postoperative pulmonary complications was lower in the PGA-glue group than in the glue-alone group (0% vs 7%, $p=0.042$) (Table 4.5) They suggested that the patients undergoing upper lobe resection and those with severe emphysema might be the best candidates for this technique.

Subcutaneous cerebrospinal fluid (CSF) leakage frequently occurs after spinal intradural surgery and is associated with the risks of poor wound healing, meningitis, and pseudomeningocele. When dural substitution is required, autografts derived from muscle and fascia can be used, but these tissues are not always available in the desired shape and quantity. Synthetic surgical membranes such as expanded polytetrafluoroethylene (ePTFE) have been used as dural substitutes, although they are associated with a high risk of postoperative CSF leakage because of the absence of adhesion to the dura mater. Sealants have been used to prevent CSF leakage from needle holes.

Fifty-three patients underwent spinal intradural surgery, and 38 procedures were performed with fibrin glue alone and 15 others with a combination of PGA fabric and fibrin glue [17]. Application of PGA fabric is shown in Fig. 4.4. After dural closure, fibrinogen solution was first rubbed into the site, and the PGA fabric was positioned. The fibrinogen and thrombin were then sprayed simultaneously.

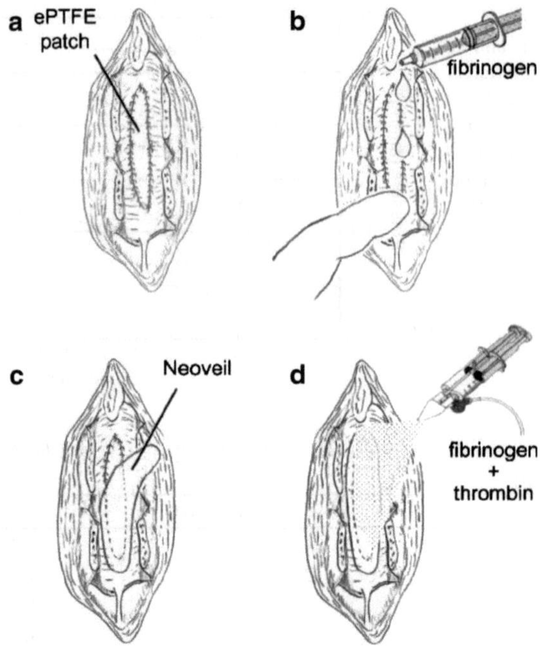

Fig. 4.4 Illustrations of the dural closure procedure performed using a poly(glycolic acid) (PGA) sheet. (**a**) The dural defect is closed using an expanded polytetrafluoroethylene (ePTFE) membrane with medium nonpenetrating clips. (**b**) After complete removal of blood and cerebrospinal fluid (CSF) from the surface of the dura mater, the fibrinogen liquid is spread manually on the dural surface. (**c**) A suitably shaped piece of Neoveil is placed on the dura mater to cover the entire dural incision with a margin of at least 3 mm. (**d**) Then, fibrinogen and thrombin liquid are sprayed simultaneously using compressed air [17]

Significantly, lower incidence of extradural and subcutaneous CSF leakage was observed in the acute (20%) and chronic (0%) stages using the PGA fabric and the fibrin glue compared to those in the acute (81%) and chronic (24%) stages, which used only the fibrin glue (Table 4.6).

The combination of PGA fabric and fibrin glue was clinically evaluated as a nonsuture dural substitute [18]. Of 160 patients, 62 patients were treated with the PGA-fibrin sheet to patch the defect, while 98 patients were treated to reinforce the fragile dura mater (Fig. 4.5). To create the PGA-fibrin sheet, the fabric was cut larger than the immediate area of the defect, and the fibrin glue was sprayed onto the fabric to the defect size to create a fibrin clot membrane supported by the mesh. Before applying the patch to the dura, the edge of the dura mater was treated with fibrinogen solution. The patch was then placed as fibrin clot side down on the defect, and the area was sprayed with fibrin glue to fix the patch to the dura mater. In 160 patients, ten (6.3%) expected subcutaneous CSF leakage. Of these, six required a second operation, and no other complications such as allergic reaction, adhesion, or infection were observed (Table 4.7). In another study, tissue replacement of the dural substitute was observed by collagenous fibers after 2 months, and there was

Table 4.6 Incidence of cerebrospinal fluid leakage [17]

	ePTFE[a] patch		No patch		Total		
	Early	Persistent	Early	Persistent	Early	Persistent	Symptomatic leakage
No PGA[a] (n=38)	86% (13/15)[b]	28% (4/14)	78% (18/23)	21% (5/23)	81% (31/38)	24% (9/37)	2.6% (1/38)[b]
PGA[a] (n=15)	25% (1/4)	0% (0/4)	18% (2/11)	0% (0/11)	20% (3/15)[c]	0% (0/15)[c]	0% (0/15)

[a]ePTFE, expanded polytetrafluoroethylene; PGA, poly(glycolic acid) sheet

[b]One patient needed a repair surgery for a cutaneous cerebrospinal fluid leak and was excluded from chronic magnetic resonance imaging study

[c]The incidence of cerebrospinal fluid leakage in the PGA group was significantly lower ($p<0.05$) than that in the No PGA group

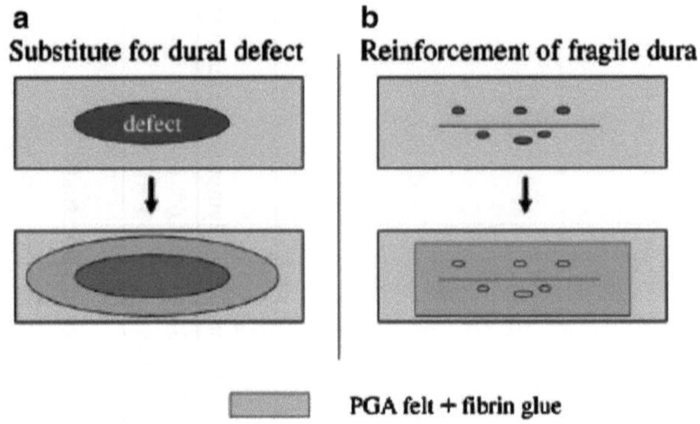

Fig. 4.5 Scheme of the two types of poly(glycolic acid) (PGA) patches used to prevent cerebrospinal fluid (CSF) leakage. (**a**) To substitute for dural defect. (**b**) To reinforce fragile dura mater [18]

Table 4.7 CSF leakage and reoperation outcomes in 160 patients who underwent spinal surgery in which the PGA sheet was used [18]

	No. of patients	CSF leakage	Reoperation
Intramedullary tumor	35	1	
Spinal schwannoma	22	2	1
Syringomyelia with Chiari type I malformation	20	3	2
Spinal AVM	18		
Spina bifida	17		
Adhesive arachnoiditis	12	1	1
Spinal meningioma	8		
Traumatic syringomyelia	7		
OPLL	5		
Chiari malformation	3		
OYL	3	2	1
Arachnoid cyst	3		
Syringobulbia	2		
Miscellaneous	5	1	1
Total	160	10(6%)	6(4%)

no evidence of adhesion to the canine brain surface [19]. It was suggested that the PGA sheet plays a different role at different stages; in the early stage after introduction, it serves as a frame to reinforce the fibrin clot membrane, while in later stages, the PGA fibers represent a cradle for collagen fiber synthesis.

Shimada et al. investigated the efficacy of PGA mesh and fibrin glue as a substitute for dural repair and compared it with fibrin glue only [20]. In a series of 20 consecutive cases, dural closure was performed by suture and fibrin glue, and in the subsequent ten cases, PGA was added to the fibrin glue. Magnetic resonance imaging (MRI) of the surgical site was retrospectively reviewed to evaluate the presence of CSF fistula or leakage after the surgery. The results showed that two patients treated with fibrin glue alone had CSF leakages due to incidental dural tears after laminectomy for ossification of the ligamentum flavum, whereas no CSF fistula was

Table 4.8 Comparison of fibrin glue and PGA mesh plus fibrin glue [20]

	No. of procedures		CSF fistula (pseudomeningocele)	
Procedure	Extradural	Intradural	Extradural	Intradural
Fibrin glue	7	13	4	1
PGA mesh plus fibrin glue	3	7	0	0

Table 4.9 Postoperative complications for fibrin glue alone (**A**) and combination of fibrin glue and PGA sheet (**B**) groups [21]

	A group ($n=37$)	B group ($n=51$)	
	No. of patients	No. of patients	p value
Postoperative bleeding	1	0	0.2
Postoperative bile leakage	3	0	0.03
Postoperative wound infection	3	4	0.3

Table 4.10 Compositions of TachoComb® fleece (1 cm² and 0.5 cm thick)

Collagen from equine tendons	2.1 mg
Human fibrinogen	5.5 mg
Human thrombin	2.0 IU
Bovine aprotinin	0.071 Ph Eur U
Riboflavin (vitamin B2) marking the coated side	7–26 μg

observed for patients treated with the PGA and fibrin glue, and neither postoperative adverse effects nor complications were observed (Table 4.8).

Bile leakage prevention by the fibrin glue and PGA mesh was investigated after hepatic resection without biliary reconstruction [21]. Of 88 patients, 37 patients were treated with the fibrin glue alone, and 51 patients were treated with the fibrin glue and PGA sheet. Postoperative complications are shown in Table 4.9. The combination of fibrin glue and PGA sheet was significantly favorable in preventing bile leakage.

Fibrin Glue Sheet

TachoComb® (CSL Behring, PA) is a ready-to-use hemostatic sheet, consisting of equine collagen fleece with surface coating of fibrin agents (human fibrinogen, bovine thrombin, and bovine aprotinin). The compositions are shown in Table 4.10. Further developments of TachoComb® are TachoComb H® and TachoSil®. TachoComb H® contains human fibrinogen, human thrombin, and bovine aprotinin, whereas TachoSil® consists of purely human coagulation agents.

The products do not require any preparation or conditioning and can be immediately applied directly to traumatized tissues, in contrast to the liquid application of fibrin glue. When the coating of collagen fleece comes in contact with fluids (i.e., normal saline, body fluid, or a bleeding surface), the components dissolve and diffuse into the cavities and begin to react. The collagen fleece helps to tamponade the wound and keeps the coagulation components in the region of bleeding. It requires 3–5 min of pressing till the area is sealed. It is mechanically stable and can be used for the treatment of diffuse bleeding of the thoracic wall. It is bioabsorbed within a few weeks. The adhesive strength is much higher than that of fibrin glue, since dried fibrinogen and thrombin will form a highly concentrated gel when a small amount of surface water is used from the site of application.

The efficacy of TachoComb H® patches in controlling suture hole bleeding was investigated in a prospective, randomized, controlled trial of 24 patients undergoing femoral anastomosis and femoral or carotid patch angioplasty with PTFE grafts [22]. The median time to hemostasis was 300 s in patients treated with TachoComb® and 600 s in the control group (surgical swabs). There were no serious complications associated with the use of TachoComb H® patches. CSF leakage occurred in only five cases. The application of TachoSil was reviewed for 408 patients with hemorrhagic risk factors or operations associated with an expected increase in bleeding [23]. The operations were performed on various organs, such as the liver, vascular system, heart, spleen, thorax, and kidney, and the results supported the efficacy and safety of TachoSil as a hemostatic agent. TachoComb® has been used as a dural substitute, and a clinical study of 288 patients revealed no superficial or deep wound infections or aseptic meningitis [24].

Gelatin-Thrombin Hemostasis

FloSeal® (Baxter Healthcare Corporation, Freemont, CA) consists of cross-linked gelatin granules and human thrombin (Thrombin-JMI®, a sterile freeze-dried powder, Jones Pharma, Inc.), which was approved by the FDA in 1999. Bovine-derived gelatin is cross-linked with glutaraldehyde and ground to 500–600 μm-sized particle. Prior to use, the thrombin is reconstituted with saline and added to the gelatin matrix component in the operating suite.

When applied to a bleeding site, the gelatin granules swell by 10–20% within 10 min, reducing blood flow and providing tamponade. It can conform to irregular wound shapes. The high concentration of thrombin reacts with fibrinogen and forms a reinforced clot around the gelatin matrix. Unreacted granules can be removed with gentle irrigation or suction. The granules in the clot are bioabsorbed by the body in 6–8 weeks. Unlike fibrin glue, this agent only works in the presence of blood.

A multicenter, randomized clinical study was performed involving 309 patients undergoing either cardiac, vascular, or spinal surgery at ten institutions [25]. They were randomized to FloSeal or the control (thrombin-soaked Gelfoam®) group. Gelfoam® is a cross-linked gelatin foam that will be discussed later. The primary effectiveness endpoint was success in hemostasis at 10 min for the first bleeding site treated. As shown in Table 4.11, FloSeal was effective within each site of application and also effective in 89% of fully heparinized patients undergoing cardiac surgery. The efficacy of FloSeal has been reported in a wide variety of surgical fields

Table 4.11 Hemostasis success at 3 and 10 min by severity of bleeding for the cardiac cohort (first treated bleeding site) [25]

	3 min		10 min	
Severity of bleeding	FloSeal	Control	FloSeal	Control
Oozing	23/35 (66%)	10/35 (29%)*	33/35 (94%)	23/35 (66%)**
Heavy bleeding	10/13 (77%)	0/10 (0%)[a]	12/13 (92%)	4/10 (40%)[b]

Fisher's exact test: FloSeal versus Control
Oozing: *at 3 min, $p=0.0037$; **at 10 min, $p=0.0057$
Heavy bleeding: [a]at 3 min, $p=0.0004$; [b]at 10 min, $p=0.0186$

including cardiac, vascular, maxillofacial and plastic facial, rhinological, urological, lacrimal, neurosurgical, thyroid, and nasal/sinus surgeries. FloSeal also decreased the tenacity of adhesion between the dura and scar at laminectomy sites in a canine model [26].

Other Types of Natural Hemostatic Agents

Use of gelatin foams as hemostatic agents was introduced by Gorrell and Wise in 1945 [27]. Gelfoam® is made of purified porcine skin gelatin foam and comes as sponge or powder. It absorbs up to 45 times its weight of whole blood and arrests bleeding by formation of an artificial clot and production of a mechanical matrix that facilitates clotting. It is bioabsorbed completely within 4–6 weeks. The powder is mixed with a sterile saline solution and applied as paste to sites of bleeding. When use of Gelfoam® in gynecologic surgery was reviewed, no excessive scar tissue was observed attributable to absorption of Gelfoam® as examined postoperatively. It should not be used in closed spaces since it swells significantly and may compress nerves.

In 2007, the Gelfoam® Plus Hemostasis kit (Baxter) was approved by the FDA. It consists of Pfizer's Gelfoam®, Baxter's Thrombin (human), and saline solution. It is indicated for use as a hemostatic device for surgical procedures when control of capillary, venous, and arteriolar bleeding by pressure, ligature, and conventional procedures is ineffective or impractical. Prior to use, thrombin is dissolved in saline solution, and Gelfoam® is soaked in it.

Oxidized cellulose was first introduced in 1942 by Frantz and launched in 1960 [27]. Surgicel® (Ethicon, Johnson & Johnson) is made of oxidized cellulose in three forms: fibrillar, knitted, and densely knitted. It has very good handling characteristics. The low pH of this material may have antimicrobial effects but also causes inflammation of the surrounding tissue and delays wound healing. Moreover, it should not be used with other hemostatic agents, such as thrombin, since its low pH will inactivate them. It should not be moistened before use, as it yields more hemostatic effect when it is dry. Although the duration of bioabsorption of this material is reported to be 7–14 days, others reported a long-term (several years) presence of Surgicel® after cardiac surgery [28].

Aviten® (C.R. Bard, Inc., NJ) consists of microfibrillar collagen, which was first developed in 1970. The large surface area of the material enables platelet adherence and activation for hemostasis. It is bioabsorbed in less than 8 weeks. Although it does not swell like Gelfoam®, it is less effective in patients with thrombocytopenia and sticks to the operator's gloves.

HemCon® and ChitoFlex® (HemCon Medical Technologies, Portland, OR) are chitosan-based hemostatic dressings. They are prepared by freeze-drying dilute aqueous acetic acid solutions of ultrapure-grade chitosan (FMC NovaMatrix, Iceland). The resultant sponges are compressed, annealed, and γ-irradiated for sterilization. It has been shown to effectively control external aggressive hemorrhage from severe traumatic injuries [29]. The feasibility of use of an internal chitosan dressing was also evaluated for control of hemorrhage and urinary leakage by sealing off parenchymal wounds following laparoscopic partial nephrectomy (LPN) in a porcine model. Use of chitosan dressing achieved immediate hemostasis and sealing of urinary leakage following laparoscopic polar resection or wedge resection of the kidney [30].

Semisynthetic Sealants

Gelatin-Based Glue

Gelatin-resorcinol-formaldehyde (GRF® glue, Cardial, Technopole, Sainte-Etienne, France) is composed of two types of solutions, GR solution (gelatin and resorcinol) and F-activator (formaldehyde and glutaraldehyde). Formaldehyde and glutaraldehyde react not only with gelatin and resorcinol, but may also react with amino groups of tissues, creating bonding between the resulting gel and the tissue. The proposed chemical structures of these components and their reaction scheme are shown in Fig. 4.6 [31, 32].

Physical properties of GRF® have been compared with cyanoacrylate gel and fibrin glues [33]. In an ex vivo study, aortic specimens from sheep were glued with warm adhesive under wet and dry conditions and applied at a defined degree of compression (5 N). When tensile strength and elasticity were assessed, the tensile strengths of GRF® glue were found to be 3.50 ± 1.64 N/cm^2 (dry) and 1.38 ± 1.00 N/cm^2 (wet), whereas those of cyanoacrylate were 4.83 ± 1.24 N/cm^2 (dry) and 3.18 ± 1.27 N/cm^2 (wet). Fibrin glue exhibited the tensile strength of 0.80 ± 0.33 N/cm^2 (dry). The bonding capacity of GRF® glue increased when pressure increased to 20 N.

Since the introduction of GRF® in 1979 for acute aortic dissection [34], it has been widely used except in the United States. In acute type A aortic dissection, the survival of patients depends on immediate diagnosis and emergent surgical intervention. Without surgery, the mortality rate increases by 1–2% per hour and is approximately 25% during the first 24 h [35, 36].

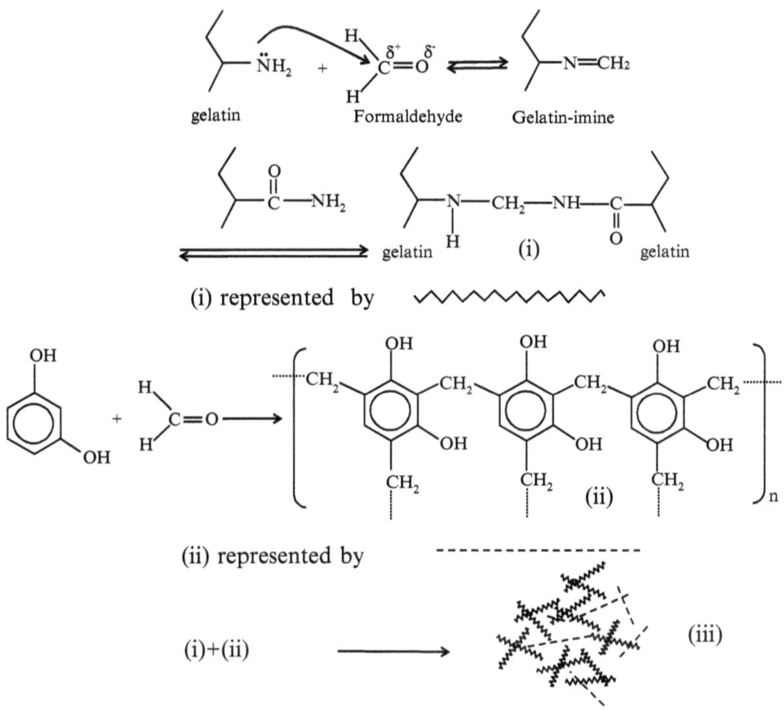

Fig. 4.6 Reaction scheme for gelatin-resorcinol-formaldehyde (GRF) glue [31]

The clinical use of GRF® is problematic due to continued controversy concerning the possible carcinogenicity and mutagenicity of formaldehyde. It has been reported that long-term exposure of formaldehyde in embalming practice and in the funeral industry is significantly associated with increased risk for mortality from myeloid leukemia [37]. Efforts have been made to eliminate formaldehyde from the composition of GRF® and to replace it with glutaraldehyde alone or glyoxal. However, these attempts were unsuccessful, as its bonding was reduced and affected the efficacy of the product, though greater durability was observed in vivo. By reducing the formaldehyde concentration and mixing with glutaraldehyde, high initial strength and increased durability were achieved [38].

Conflicting data are found in the literature on the effect of GRF® glue in aortic surgery. Long-term stability of the dissected, GRF®-treated aorta has been reported, whereas other studies found late adverse events such as late reoperation and distal embolization [39–41]. Suzuki and colleagues reviewed 269 patients (222 patients with GRF® applied and 47 patients without GRF®) who underwent emergency surgery for acute type A aortic dissection [42]. The rate of inhospital mortality was significantly higher in the non-GRF® group (31.9%) than in the GRF® group (12.6%), indicating that the short-term outcome of surgery was improved with the use of GRF®. However, false aneurysms were found in 31 patients (16%) 1–65 months postoperatively. Reoperation was carried out for 24 of these patients, and it was found that the site of GRF® glue application had degenerated with a widely opened anastomosis between the aortic root and prosthesis. Microscopic examination revealed necrosis of medial smooth muscle cells. Bingly and colleagues reviewed 67 cases of tissue glue use, with the majority of operations for type A dissection (76%), to identify any potentially glue-related complications [43]. There were two intraoperative deaths; 27 of 65 patients (41%) required 29 further open-chest operations, and 12 patients (18%) underwent late reoperation months to years later. These complications were related to the application of GRF®. Hata and colleagues reviewed 56 patients undergoing emergency surgery for acute type A aortic dissection and applied GRF® [44]. Adverse outcomes included bleeding requiring reoperation in seven patients (10%), stroke in five patients (7%), and early mortality in eight patients (12%). As these differences may arise from excessive use of F-activator, Kunihara et al. [45] used GRF® glue with a minimal amount of F-activator for 21 of 49 patients undergoing Stanford type A acute aortic dissection within 48 h with outer felt reinforcement. When applied properly, GRF® was clinically safe with regard to midterm death or reoperation, though combined use of felt reinforcement was recommended.

Albumin-Based Glue

In BioGlue® (CryoLife, Inc., Kennesaw, Georgia), stoichiometrically equivalent doses of bovine serum albumin (45% solution) and glutaraldehyde (10% solution) are mixed in a custom cartridge delivery system (Fig. 4.7) [46]. It was approved by the US FDA in 1999 for use during surgical repair of acute thoracic aortic dissection and in 2001 for general use as a hemostatic adjunct in cardiac and vascular surgery. Examples of applications are shown in Fig. 4.8 [47]. It sets as a gel within 20–30 s and establishes a firm adhesive bond due to the reactive glutaraldehyde which covalently cross-links the albumin molecules to each other as well as to the tissue proteins at the site of application. The reaction scheme is shown in Fig. 4.9.

In comparison to fibrin glue, BioGlue® has high adhesive strength to tissues and is much stiffer. The material properties of BioGlue®, PEG-based gel (CoSeal®), and fibrin glues (Tisseel™ and Crosseal®) were investigated [48]. Mean elastic moduli varied considerably among these glues: BioGlue® (3,122 ± 1,640 kPa), CoSeal® (100 ± 68 kPa), Tisseel™ (103 ± 41 kPa), and Crosseal® (54 ± 33 kPa). Clearly, BioGlue®

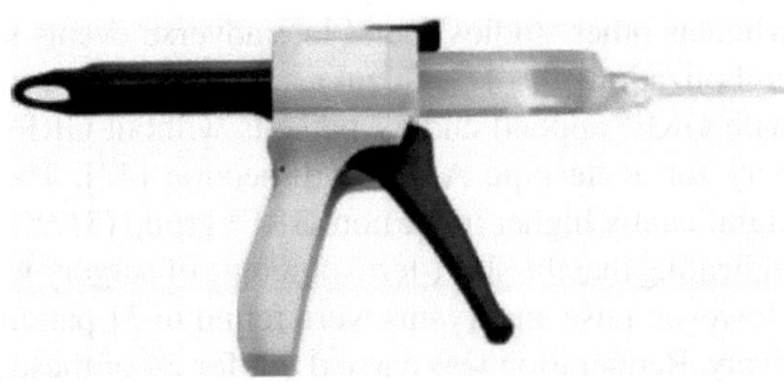

Fig. 4.7 BioGlue® applicator [46]

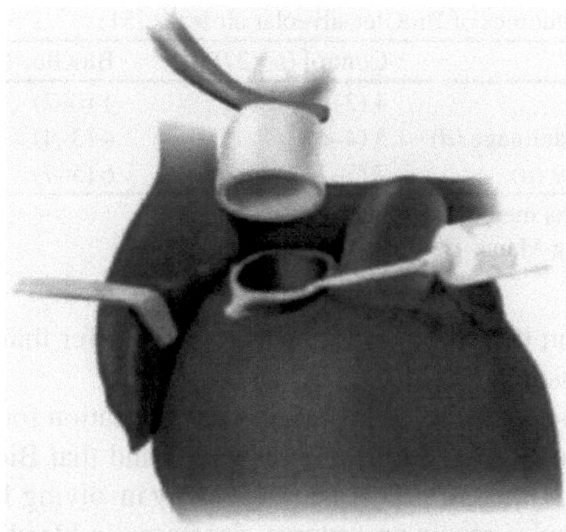

Fig. 4.8 BioGlue® being used to reinforce the repair of an aortic dissection [47]

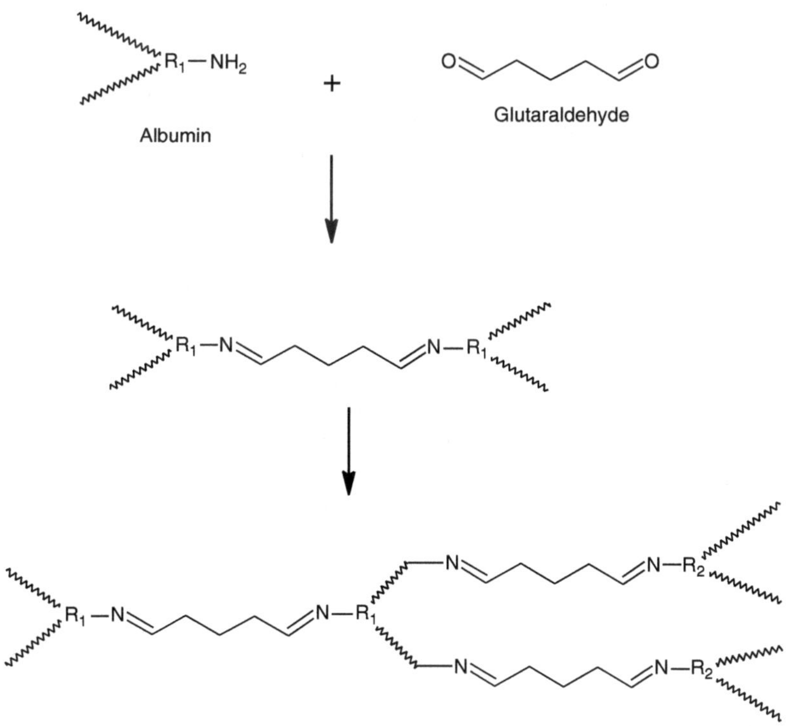

Fig. 4.9 Reaction scheme for BioGlue®

Table 4.12 Primary outcomes of BioGlue alveolar air leaks [51]

	Control ($n=27$)	BioGlue ($n=25$)	p value*
Duration of air leak (*d*)	4 (2–6)	1 (0–2)	<0.001
Duration of intercostal drainage (*d*)	5 (4–6)	4 (3–4)	0.012
Duration of hospital stay (*d*)	7 (7–10)	6 (5–7)	0.004

All numbers presented as median with interquartile range.
*p values obtained using Mann–Whitney test

was much stiffer than the other glues and was even stiffer than both Dacron grafts and human aortic tissue.

In 2007, Zehr reported detailed techniques of application for the use of BioGlue® in a variety of cardiovascular surgical cases and found that BioGlue® was safe and effective when used properly [49]. In a clinical study involving 151 patients undergoing cardiac and vascular repair procedures, anastomotic bleeding was significantly reduced in the BioGlue® group (18.8%, $n=76$) compared with the control group (42.9%, $p<0.001$, $n=75$). The use of pledgets reduced bleeding in the BioGlue® group (26.2%) compared with the control group (35.9%, $p=0.047$) [50]. In application in thoracic surgery, a randomized, controlled trial of the effectiveness of BioGlue® was carried out in 52 patients being treated for alveolar air leaks [51]. The results of duration of air leakage, intercostal drainage, and hospital stay are summarized in Table 4.12. Surgical treatment with BioGlue® exhibited statistically significant effectiveness in all categories compared to the control group (surgical treatment only).

In an animal study, BioGlue® was used on tracheal resection in the rabbit, a sensitive model for investigation of the biocompatibility of glues. Tracheal anastomosis was performed with a maximum of four sutures, and the line of anastomosis was then circumferentially covered with the adhesive. In control animals, two interrupted, airtight running sutures were placed. Over up to 12 weeks of macroscopic observation, glued tracheal anastomoses were tightly sealed, suggesting sufficient tensile strength under physiological conditions. Microscopically, inflammatory response and numerous multinuclear giant cells were present at 4 weeks but were decreased after 12 weeks.

Despite its success in clinical studies, the toxicity of unreacted and leached glutaraldehyde and the dense postoperative gel structure have been the concerns and have limited the application of BioGlue®. Fürst et al. investigated the release of glutaraldehyde from BioGlue® and its toxicity in vitro and in vivo [52]. It was found that 100–200 μg/mL of glutaraldehyde was released from 1 mL of gelled BioGlue® into 5 mL of saline and that the supernatant exhibited cytotoxicity in cultured human embryonic fibroblasts (MRC5) and mouse myoblasts (C1C12). An animal study using rabbits showed that the application of BioGlue® had serious adverse effects on lung and liver tissues, while aortic tissues exhibited low- to medium-grade inflammation. The reason suggested for the high resistance of aortic tissues was that the adventitial layer was less damaged during clinical application on aortic dissections than on raw tissue surfaces of lung and liver, since the aortic tissue is composed of large amounts of ECM with low cell content. Other in vivo studies also revealed toxic effects of BioGlue® in lung and nerve tissues and little induction of inflammation in the aorta. A high infection rate (13%) was observed in cases of pediatric craniotomy and laminectomy when BioGlue® was used [53]. It was suggested that intensive acute and chronic inflammatory response due to the BioGlue® caused expansion and weakening of the surrounding tissue, leading to secondary infection once the incision broke down.

Synthetic Sealants

Cyanoacrylates

Cyanoacrylates are monomers containing esters of cyanoacrylic acid. They were first described in 1949 [54] and have been used clinically since the late 1950s. Various homologues of cyanoacrylate adhesive have been studied including methyl, ethyl, isobutyl, isohexyl, and octyl cyanoaclylates. The structure of methyl cyanoaclylate is shown in Fig. 4.10. These monomers rapidly polymerize when in contact with water or hydroxyl groups on the actual surface being glued [55]. These create strong and flexible films as sealants to bond apposed wound edges. Use of these tissue glues is faster and less painful than that of suturing.

The shorter chain esters may be more toxic either directly or through their degradation products, which are cyanacrylates and formaldehyde. The degradation products accumulate in tissues and produce acute and chronic inflammation. In contrast, longer alkyl chains (e.g., ten carbons) are less toxic. The potential toxicity of cyanoacrylates within patients as well as among health professionals who are occupationally exposed to cyanoacrylates has been investigated. The results suggest that cyanoacrylates are dangerous even to medical staff, who are instructed to avoid direct contact, and should be used in a well-ventilated room.

Butyl cyanoacrylate has been used only for small low-tension lacerations and incisions due to its poor tensile strength and brittleness, whereas octyl cyanoacrylate yields stronger and more flexible adhesives [56]. In an animal model, various commercially available cyanoacrylate products have been compared for wound

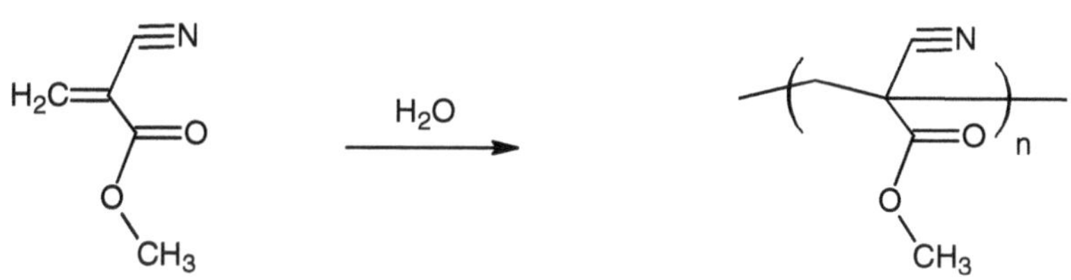

Fig. 4.10 Cyanoacrylate monomer and polymer

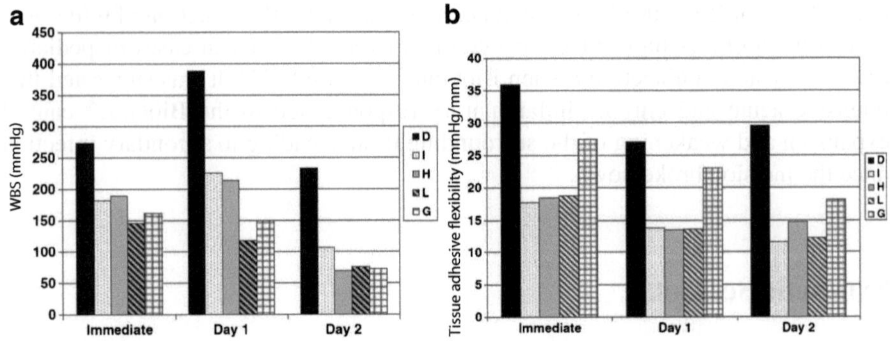

Fig. 4.11 (**a**) Wound-bursting strength (WBS) (mmHg) and (**b**) tissue adhesive flexibility (mmHg/mm) of commercial cyanoacrylate products. *D* Dermabond, *I* Indermil, *H* Hystoacryl, *L* Liquiband, *G* GluStich [57]

Table 4.13 Examples of surgical and medical applications of cyanoacrylates [58]

Specialty	Past, present, and potential applications
General surgery	Surgical wound repair Control of hemorrhage
Emergency medicine and general practice	Traumatic wound repair
Endoscopy	Control of variceal bleeding and obliteration of esophagogastric varices
Ophthalmology	Temporary repair of corneal perforations
Thoracic surgery	Closure of pulmonary leaks
Otological surgery	Repair of peripheral nerves
Interventional radiology and cardiology	Ossicular chain reconstruction
Pediatrics	Embolotherapy of various vascular abnormalities, including aneurysms
Pediatric endoscopic surgery	Wound closure in children
Pharmacotherapeutics	Tissue approximation and hemostasis Drug carrier

bursting strengths immediately after closure and 1 and 2 days after [57]. Dermabond (Colure Medical Corporation, Raleigh, NC), which consists of octyl-2-cyanoacrylate, was significantly stronger and more flexible than all the cyanoacrylate-based adhesives tested at all time points of measurement (Fig. 4.11).

The skin edges are held in apposition, and cyanoacrylate is applied in layers along the entire length of the wound. Spontaneous release of heat occurs as the polymer forms. The agent remains adherent to the skin for approximately 7–10 days while the wound heals, and comes off as the superficial layers of skin exfoliate.

Cyanoacrylates have been used mainly for external surfaces as tissue adhesives for surgical and traumatic wound repair, due to their toxicity. Examples of surgical and medical applications of cyanoacrylates are shown in Table 4.13 [58]. Surface application of Dermabond has been shown to be very successful. A randomized, control trial showed that the wounds closed by Dermabond had no noticeable differences in subjective cosmetic results and were associated with complications at 3 months postoperatively, compared with wounds closed using monofilament sutures. Adhesives also had the advantage of much faster operative time than traditional suturing.

To identify the rate of infection associated with spinal surgery after wound closure with octyl-2-cyanoacrylate, a study was performed in a total of 235 patients with one- or two-level surgery of the cervical or lumbar spine [59]. Skin closure was performed with octyl-2-cyanoacrylate after anterior cervical ($n=100$) and dorsal lumbar ($n=135$) spinal surgery. Results were compared with the infection rates of similar surgeries in the literature and those for a previous group of 503 patients who underwent wound closure with standard skin sutures after spine surgery. Only one patient suffered from postoperative wound infection, indicating an infection rate of 0.43%. The infection rates in the literature and the control group with suturing after spine surgery were 3.2% and 2.2%, respectively. No risk factor was identified.

Another widely used cyanoacrylate product, Histoacryl (B Braun, GmbH, Tuttlingen, Germany), consists of n-butyl-2-cyanoacrylate monomer. It is supplied in 0.5-ml single-use plastic ampoules, but is not approved in the USA. Clinical studies have shown its effectiveness for skin wound closure [60–62] as well as in sclerosation therapy for esophageal and fundal varices [63–65].

PEG-Based Sealants

Poly(ethylene glycol) (PEG)–based sealants are flexible and firmly adherent even on wet tissues. These hydrogels also have the ability to prevent adhesions. DuraSeal™ (Confluent Surgical Inc., Waltham, MA) is one such product developed for watertight closure after dural suturing. Two solutions, PEG-ester and trilysine amine, are mixed together and sprayed on the site of treatment. The ester-end-modified PEG and trilysine rapidly react via electrophilic nucleation reaction and form an elastic hydrogel (Fig. 4.12). One solution also contains blue dye for visual inspection. PEG has been widely used medically and has been proven safety. Trilysine is a synthetic product derived from three molecules of L-lysine, an essential amino acid. The gel is bioabsorbed in 4–8 weeks, and the product of breakdown of PEG is rapidly bioabsorbed into the bloodstream and cleared primarily via glomerular filtration in the kidneys [66]. The sealant was approved for use as an adjunct to suturing of dural repairs during cranial surgery by the FDA and the EC in 2005 and 2003, respectively. It was also approved by the EC for use as a cranial, spinal, thoracic, and vascular sealant. When applied, the solutions diffuse into pores on the tissue surface, and the subsequent polymerization results in interlocking with the roughness of the tissue surface.

In a clinical study, the efficacy of DuraSeal™ was investigated as an adjunct to sutured dural closure in reducing the incidence of postoperative incisional CSF leakage after posterior fossa surgery [67]. The results of clinical outcomes comparing DuraSeal™ with the control group using fibrin glue are shown in Table 4.14. Only 2%

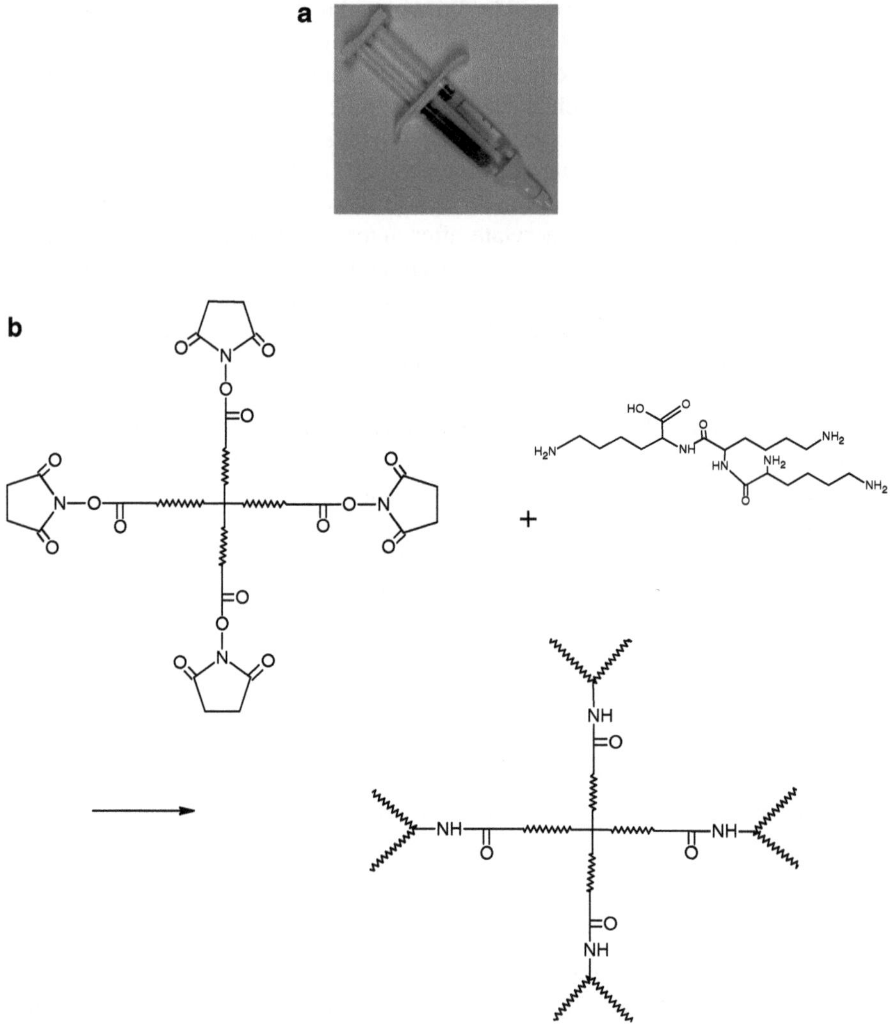

Fig. 4.12 (**a**) DuraSeal™ applicator and (**b**) reaction scheme for DuraSeal™

(2 of 100) of patients developed postoperative CSF leakage in the DuraSeal™ group, whereas 10% (10 of 100) of patients in the fibrin glue group had incisional CSF leakage ($p = 0.03$).

Some complications associated with large swelling of DuraSeal™ have been reported. Two case reports illustrated patients who developed quadriparesis after surgery as a result of expansion of hydrogel [68]. The manufacturer states the following: "Do not apply to confined bony structures where nerves are present since neural compression may result due to hydrogel swelling." It has been reported that

Table 4.14 Results of the univariate analysis of the clinical outcomes comparing the PEG dural sealant to the control group of fibrin glue–augmented dural closure[a] [67]

Clinical outcomes	PEG sealant group ($n=100$)	Control group ($n=100$)	Odds ratio (95% CI)	p value
Incisional CSF leak	2	10	0.18 (0.039–0.86)	0.03[c]
Time to leak (d)	28±14	9.1±5.3	n/a	0.005[c]
Pseudomeningocele	8	5	1.7 (0.52–5.2)	0.4
Wound infection	4	2	2.0 (0.37–11.4)	0.7
Meningitis	2	5	0.39 (0.07–2.0)	0.4
Length of stay (d)	5±3.6[b]	6.3±4.3	n/a	0.02[c]

[a]*CI* confidence interval, *PEG* poly(ethylene glycol), *CSF* cerebrospinal fluid
[b]These data exclude one patient whose length of stay (42 day) was considered to be an outlier
[c]Statistically significant

the basic spray applicator provided in the DuraSeal™ kit may not provide adequate control in limiting thickness, resulting in large amounts and increased thickness of hydrogel [69]. An air-assisted MicroMyst Applicator was developed to control the volume and thickness of hydrogel. Another study reported DuraSeal™-entrapped hematoma as a complication [70]. After surgery, DuraSeal™ was found to form a restrictive layer, completely entrapping the extradural hematoma with resultant spinal cord compression. It was suggested that its use be limited to thin-layer application to the defective area only to avoid this complication.

To overcome the problem of swelling after surgery, DuraSeal™ has been modified to increase cross-linking density, and the new product is called DuraSeal Xact™ and has been approved by the EC. This product swells very little or not at all after application. Due to the increase in cross-linking density, degradation times are increased to 2–3 months. The efficacy of DuraSeal Xact™ was investigated in a canine laminectomy model of CSF leakage and adhesion formation [71]. After full-width L-2 and L-5 laminectomies, 1-cm midline durotomies were created and closed by suturing except for the last 1–2 mm on the cranial end to create spontaneous CSF leakage. DuraSeal Xact™ was applied either with a conventional 2-chamber syringe sprayer (Dual Liquid applicator) or air-assisted MicroMyst applicator. The control group did not receive any further treatment. At 2 and 4 months after surgery, animals were sacrificed and evaluated for dural healing and epidural adhesion formation. Figure 4.13 shows the percentage of durotomy suture lines leaking CSF at the completion of surgery (intraoperatively) and the percentage of animals with fluid accumulation after surgery (postoperative). Application of hydrogel effectively prevented CSF leakage. The use of MicroMyst helped to control application, and the hydrogel formed was significantly less thick. Adhesion was significantly reduced by gel application compared to the control group, which exhibited extensive, dense epidural adhesions. At 4 months, the hydrogel was completely absorbed, and the space was filled with scant, loosely organized connective tissue.

CoSeal® (Baxter, Fremont, CA), approved by the FDA in 2001, is another PEG-based hydrogel that is used in vascular reconstructions to achieve adjunctive hemostasis by mechanically sealing the region of leakage. The reaction scheme is

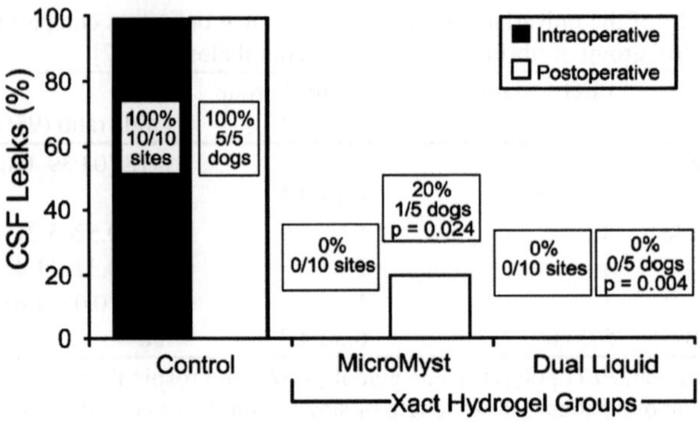

Fig. 4.13 Percentage of durotomy suture lines leaking cerebrospinal fluid (CSF) at the completion of surgery (intraoperative) and the percentage of animals with postoperative subcutaneous fluid accumulations (presumed to be CSF) [71]

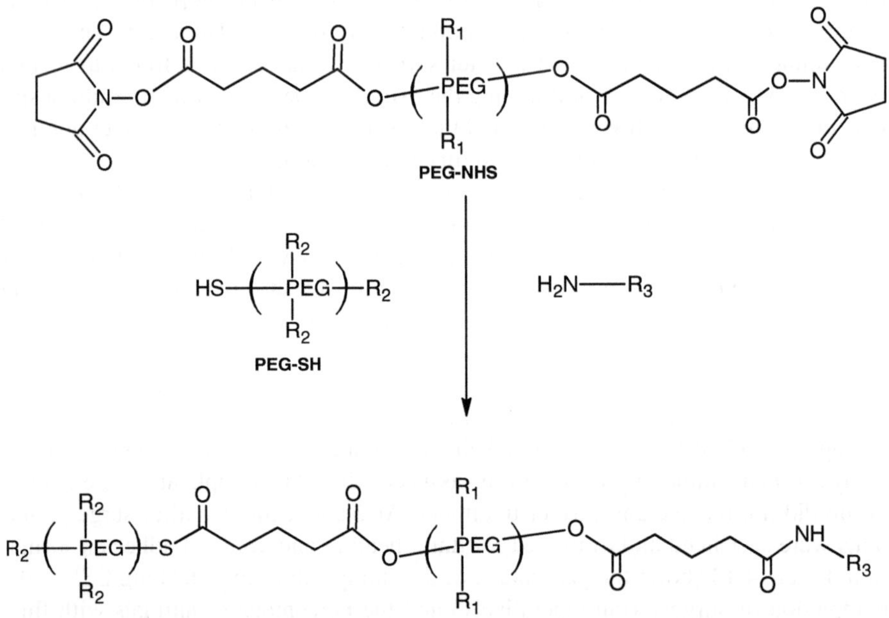

Fig. 4.14 CoSeal® reaction. Tetrafunctional succinimdylglutaryl poly(ethylene glycol) (PEG) macromer reacts with a thiol PEG macromer (*left*) and protein (*right*) to form a dimer, which leads to the formation of cross-linked polymers

shown in Fig. 4.14. It polymerizes in ~5 s. The hydrogel swells approximately 400% over 24 h. A randomized, controlled trial was performed to evaluate the efficacy of CoSeal® for anastomotic suture hole bleeding and compare it to Gelfoam®/Thrombin [72]. One hundred forty-eight patients scheduled for implantation of PTFE grafts were randomly treated with either CoSeal® ($n=74$) or control ($n=74$).

Table 4.15 Anastomotic sealing success of CoSeal® and control (Gelfoam®/Thrombin)[a] [72]

	CoSeal®	Control	p*
10-min sealing			
Overall	64/74 (86)	59/74 (80)	0.29
Infrainguinal bypass graft	23/29 (79)	19/26 (73)	0.62
Dialysis access shunt	40/43 (93)	37/44 (84)	0.14
Immediate sealing			
Overall	35/74 (47)	15/74 (20)	<0.001
Infrainguinal bypass graft	13/29 (45)	3/26 (12)	0.003
Dialysis access shunt	22/43 (51)	10/44 (23)	0.002

[a]Data are given as number (percentage) unless otherwise indicated
*Significance levels computed with the Cochran-Mantel-Haenszel χ^2 test for sealing success rates

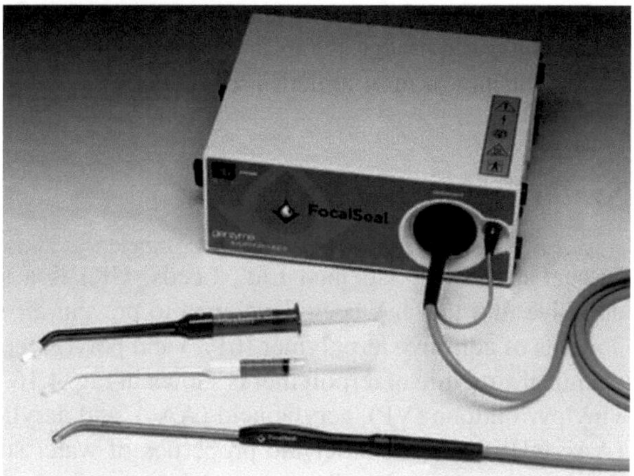

Fig. 4.15 FocalSeal delivery system [74]

The success in anastomotic sealing is summarized in Table 4.15. The efficacy of CoSeal® was equivalent to that of Gelfoam®/Thrombin.

In vitro testing of DuraSeal™ and CoSeal® was performed in comparison with fibrin sealant to evaluate properties such as preparation time, mechanical strength, water uptake, and burst pressure strength [73]. Testing was performed immediately after preparation and after soaking in PBS at 37°C for 1, 2, and 3 days. The preparation times for DuraSeal™ and CoSeal® were 1.0 ± 0.0 min and 3.2 ± 0.0 min, respectively, and were faster than that of fibrin sealant (12.3 ± 2.0 min). Over a 3-day period, CoSeal® swelled about 560% by weight, whereas DuraSeal™ increased about 100%, and fibrin sealant by about 3%. The strength of interface bonding between collagenous materials and the sealant was determined, and it was found that the burst pressure of DuraSeal™ was significantly higher than those of CoSeal® and fibrin sealant.

FocalSeal (Genzyme Biosurgery, Inc., Cambridge, MA), which is also known as AdvaSeal, is another PEG-based product. It is derived from photo-polymerization of PEG and oligotriethylene carbonate with acrylate ester end caps, in the presence of triethanolamine and eosin Y as a photo-initiator, using visible light from a xenon arc lamp (Fig. 4.15) [74]. It has been clinically used for sealing of pulmonary air

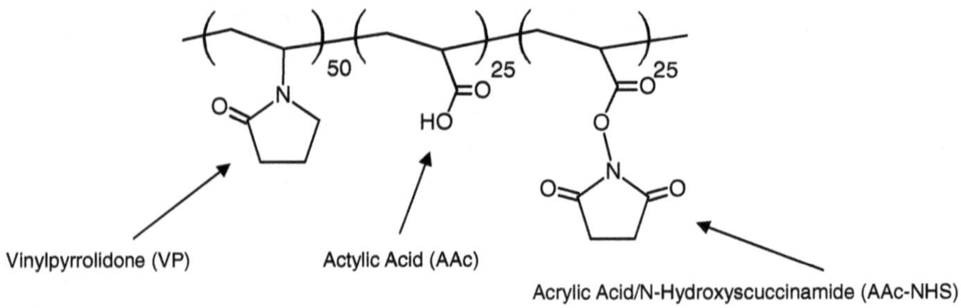

Fig. 4.16 TissuePatch3™ [77]

leakage [75]. Tanaka et al. investigated the sealing of false channels in acute aortic dissection using mongrel dogs and found that the sealant was effective for hemostasis and closure of false lumens [76].

Synthetic Adhesive Sheet

TissuePatch3™ (Tissuemed Ltd., Leeds, UK) is a synthetic, bioabsorbable self-adhesive film that has been developed to prevent air and liquid leaks in surgery. It consists of adhesive terpolymer (61%) and poly(glycolide-co-lactide) (33.5%). The chemical structure of terpolymer is shown in Fig. 4.16. It contains three components: vinylpyrrolidone (VP), acrylic acid (AAc), and acrylic acid/*N*-hydroxysuccinimide (AAc-NHS), which confer the properties of water solubility, electrostatic interaction, and covalent bonding potential, respectively [77]. No preparatory step is required, and it adheres within 30 s of application. The efficacy of TissuePatch3 was clinically evaluated for sealing of air leakage after lung surgery [78]. It was found that 80% of patients (12 of 15 patients) became air leakage–free at the end of the surgical procedure. No device-related adverse events were observed.

Recent Developments

Albumin-Based Sealant

Albumin-based hydrogel sealant (ABHS), which is based on human serum albumin and PEG disuccinimidyl succinate as a cross-linker, was investigated for prevention of air leakage in a rat model [79]. The chemical structures of its materials are shown in Fig. 4.17. Upon mixing of the two components, cross-linking occurred within 15 s, due to the formation of amide bonds between NHS end groups of PEG and free amino groups of albumin. An air leakage model was created in rat lung by carefully

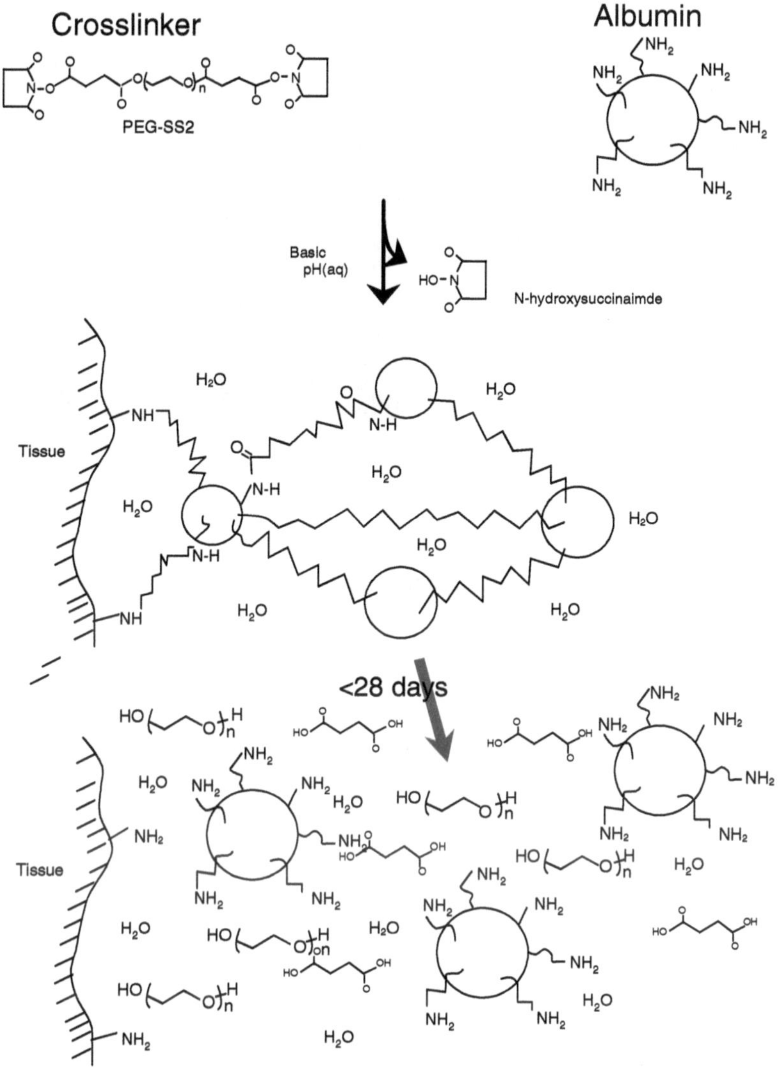

Fig. 4.17 Chemistry of starting materials and albumin-based hydrogel sealant (ABHS) [79]

making incisions of 6 mm in length and 2 mm in depth and measuring the burst pressures of adhesives. As shown in Fig. 4.18, ABHS had significantly higher initial burst pressure (i.e., 77.5 ± 19.1 mm Hg) than fibrin sealant (i.e., 30.8 ± 15.2 mm Hg), whereas no significant difference was observed between the two groups at day 7.

Taguchi et al. developed biological glues combining an organic acid derivative and proteins such as collagen, gelatin, and human serum albumin (HSA). The active ester groups of the organic acid derivative react with amino groups of not only proteins of glue but also applied tissue surfaces. Citric acid derivative (CAD) was used

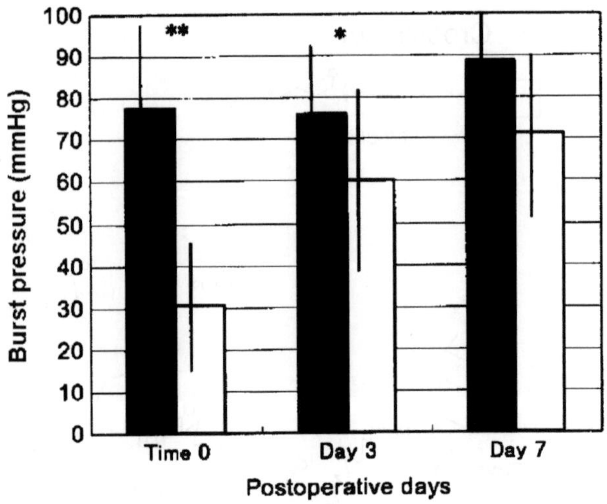

Fig. 4.18 Results of the burst pressure test for all the cases. *Solid bars*, albumin-based hydrogel sealant; open bars, fibrin glue. *Statistically significant difference ($p<0.05$). **Statistically significant difference ($p<0.001$) [79]

as a cross-linking agent. Citric acid, which is the source of CAD, is metabolized in the tricarboxylic acid cycle (TCA cycle) in the mitochondria. Although gels consisting of CAD and collagen and CAD and HSA have exhibited sufficient bonding strength in vitro and insignificant toxicity, both required comparatively long periods of time to reach a certain bonding strength. To reduce the bonding time, tartaric acid derivative (TAD) was used as a cross-linking agent, and the bonding strengths of glues (TAD-A) developed by reacting different concentrations of TAD (0.1, 0.3, and 0.5 mmol) and HSA (44 w/w%) were compared with those of fibrin glue and GRF® [80]. On the backs of mice, full-thickness skin incisions of 2 cm were made with surgical scissors, and wounds were closed immediately with 1-mg portions of glues. Wounds were then held together for 5 min. Three hours after glue application, animals were sacrificed, and skin was obtained and trimmed in rectangular fashion. A uniaxial tensile machine was used to measure the bonding strength, and the results are shown in Fig. 4.19.

The bonding strengths of glues increased with increasing TAD concentration, and even the lowest TAD glue exhibited significantly higher bonding strength than fibrin glue. The bonding strengths of glues with 0.3 and 0.5 mmol TAD were significantly higher than that of GRF®. Bonding strengths of tissues gradually decreased until 7 days and then increased after 14 days. This decrease indicates the bioabsorption of glue and broken extracellular matrix (ECM), while the increase in bonding strength demonstrates remodeling of the subcutaneous tissue structure. When the glue was implanted subcutaneously in mice to investigate the inflammatory reaction in surrounding tissues, the glue with low TAD concentrations (0.1 and 0.2 mmol) exhibited very mild inflammation, whereas the 0.3 and 0.5 mmol TAD gels were associated with moderate inflammation, possibly due to the local acidification by *N*-hydroxysuccinimide, which separates from carboxyl groups after the cross-linking reaction.

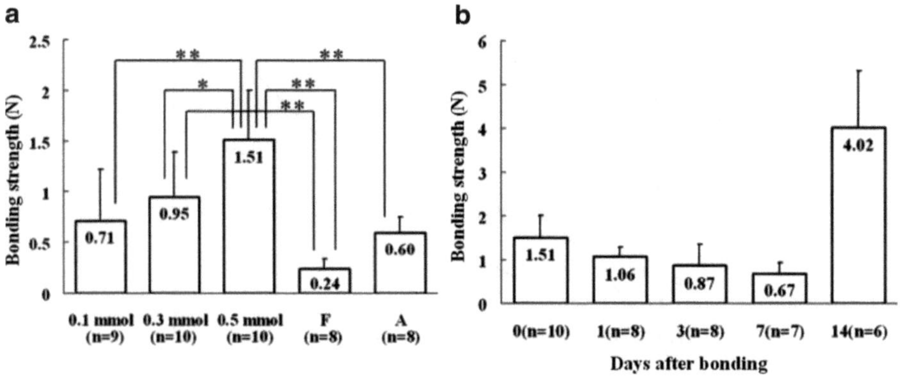

Fig. 4.19 (**a**) Maximum bonding strength of the skin bonded with tartaric acid derivative (TAD-A) glue with different TAD concentrations and commercial biological glues, 3 h after application. *F* fibrin glue (Bolheal), *A* aldehyde glue (GRF® glue). (**b**) Results of serial changes in bonding strength for the TAD 0.5 mmol group [80]

Fibrin-Based Sealant

Fibrinogen can be rapidly photo-cross-linked in the presence of ruthenium as the trisbipyridyl complex and the oxidant ammonium or sodium persulfate by light to yield high-molecular-weight products that have strong in vitro adhesive properties [81]. Fancy and Kodadek [82] first described this technique, which results in a safe and effective method of cross-linking a natural matrix protein to endogenous tissue. This method is independent of thrombin, calcium ions, and other components of the coagulation cascade. A mixture of fibrinogen (150 mg/mL), 2 mM $Ru^{II}(bpy)_3Cl_2$, and 20 mM sodium persulfate (SPS) was prepared and applied immediately to wounds, followed by illumination with a 600 W tungsten-halide light source at 150 mm for 20 s [81]. The fibrinogen associated with endogenous fibrinogen as well as other major protein components of the ECM to enable subsequent formation of $Ru^{II}(bpy)_3Cl_2$ – catalyzed covalent bonds. Intermolecular protein associations account for the superior adhesive properties of the photo-cross-linked fibrinogen tissue sealant. The sealant was well tolerated and persisted up to 8 weeks but completely dissolved by 18 weeks with minimal inflammatory response. Repair of skin incisions in rats or use as an arterial hemostat in pig arrested bleeding within 20 s of application.

Gelatin-Based Sealant

Different cross-linking types of gelatin-based glues have been developed. One of such is the sealant consisting of gelatin and glutaraldehyde (GA) as a cross-linking agent [83]. The glue consisting of 23 wt.% gelatin and 0.12 wt.% GA exhibited

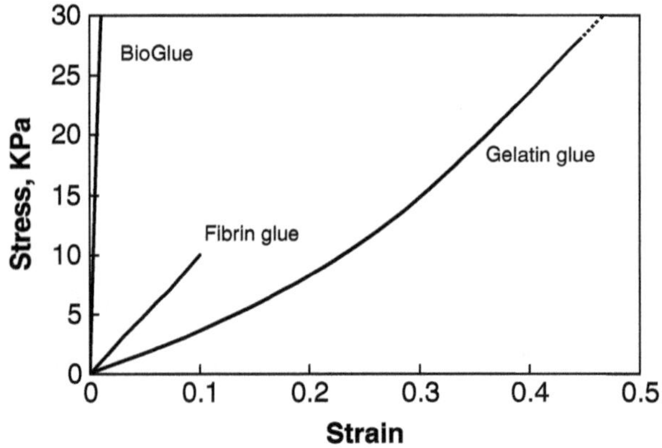

Fig. 4.20 The stress–strain relationship for BioGlue, fibrin glue, and gelatin glue [84]

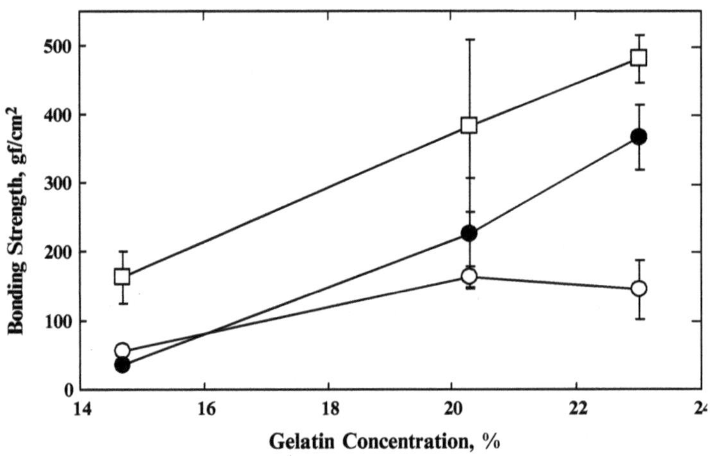

Fig. 4.21 Influence of gelatin and GA concentrations on the bonding strength of gelatin glue to chicken leg tissue. 0.12 wt.% (*open circle*), 0.24 wt.% (*closed circle*), and 0.59 wt.% (*square*) GA in gelatin glue [83]

more than 90% degradation in 7 days when implanted subcutaneously in rats. Figure 4.20 shows the stress–strain relationship of BioGlue®, fibrin glue, and gelatin glue [84]. All samples were tested in their hydrated states. Gelatin glue was found to have high elasticity and high strength. The bonding strength between this glue and two pieces of chicken tissue was compared to that of commercial fibrin glue (Beriplast®) [83]. As shown in Fig. 4.21, the bonding strengths of gelatin glue increased with increasing gelatin and GA concentrations. The gelatin glue consisting of 23 wt.% gelatin and 0.12 wt.% GA showed the bonding strength of 144 ± 44 gf/cm^2, whereas that of fibrin glue was 48 ± 20 gf/cm^2. When the GA concentration was raised to 0.59 wt.%, the bonding strength increased to 480 ± 35 gf/cm^2, and even the

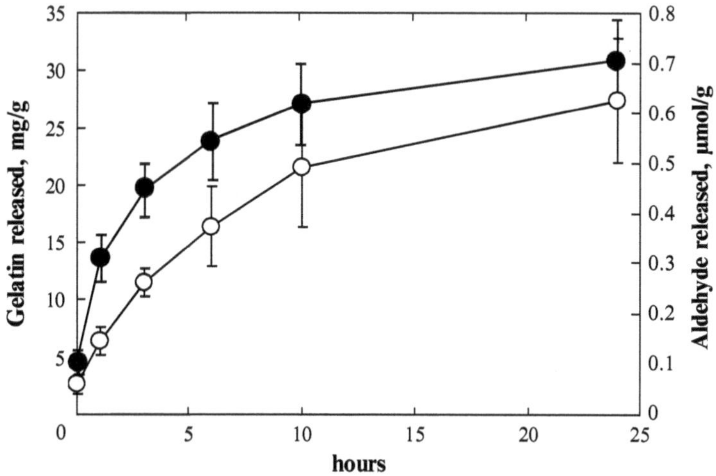

Fig. 4.22 Time course of gelatin (*open circle*) and aldehyde (*closed circle*) release from the 1 g of wet gelatin glue (final gelatin and GA concentrations of 23 and 0.12 wt.%, respectively) [83]

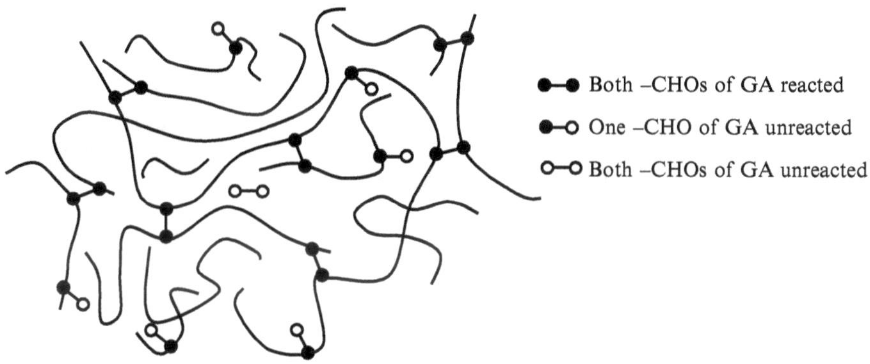

Fig. 4.23 Reaction of –CHO groups of GA with gelatin molecules having various chain lengths

tissue broke off in some cases. The release profiles of aldehyde and gelatin from the glue consisting of 23 wt.% gelatin and 0.12 wt.% GA in PBS (–) at 37°C are shown in Fig. 4.22. It was found that 0.7 ± 0.1 μmol/g of aldehydes was released in 24 h. The molar ratio of $-NH_2$/–CHO group in this glue is calculated to be approximately 7/1. This high ratio suggests that the most GA would react with gelatin molecules, as schematically shown in Fig. 4.23, and the extracted aldehydes are mostly from the low-molecular-weight gelatin molecules having an unreacted –CHO group.

One broad type of application of sealant is to close needle holes at the site of anastomosis of vascular grafts. The glue consisting of 23 wt.% gelatin and 0.12 wt.% GA was applied with or without rubbing in with fingers to seal the needle hole on the PTFE vascular graft (Gore-Tex). The water pressure in the graft was increased,

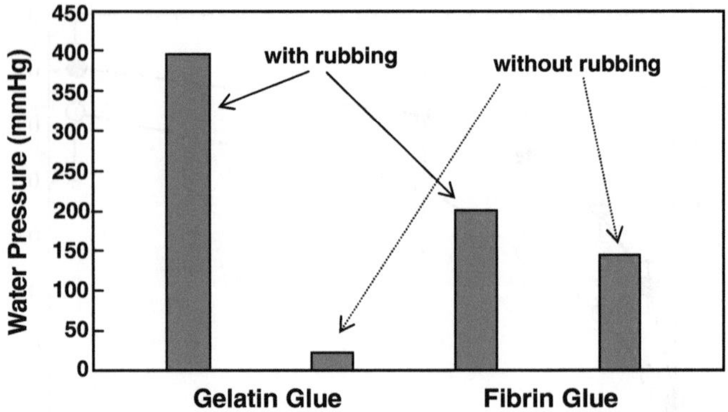

Fig. 4.24 Burst water pressure of needle hole on PTFE vascular prosthesis sealed with gelatin glue prepared at final concentrations of 23 wt.% gelatin and 0.12 wt.% GA or fibrin glue applied with or without rubbing with fingers [83]

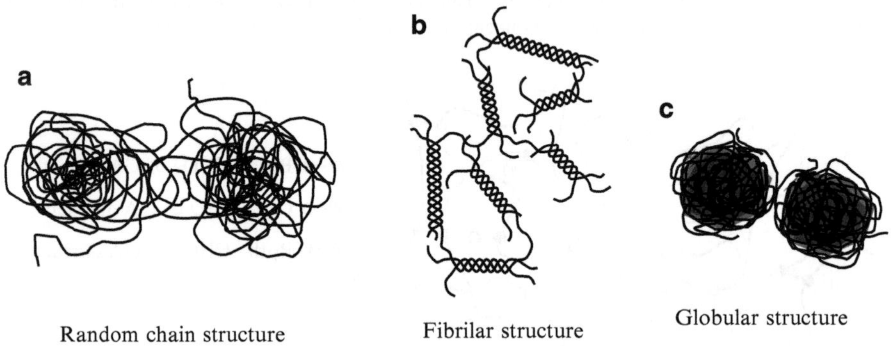

Fig. 4.25 Three different chain structures in solution

and the burst pressure was measured when water leakage occurred (Fig. 4.24). The burst pressure of the gelatin glue applied with rubbing was found to be 400 mmHg, while that of fibrin glue was 200 mmHg. The high viscosity of gelatin solution also prevents leakage through suture-line needle holes, which is often a problem with the low viscosity of serum albumin in BioGlue® [85]. The efficacy of gelatin glue was investigated as a sealant for air leakage from lungs; it was found to be significantly more effective than fibrin glue. The superiority of gelatin glue compared to those of fibrin and albumin glues comes from the structural differences of these protein molecules. As represented in Fig. 4.25, gelatin is present in solution mostly as random chains, whereas fibrin and serum albumin are fibrillar and globular structures, respectively. High viscosity and strong adhesion of gelatin glue are due to the chain entanglement that is possible due to the random chain structures. Furthermore, gelatin is partially under triple helix conformation, which produces a strong gel.

A rapidly curable, bioabsorbable glue consisting of gelatin and poly(L-glutamic acid) (PLGA) was developed. [86, 87]. Mixing of gelatin and PLGA aqueous solutions in addition to water-soluble carbodiimide (WSC) forms a hydrogel within several

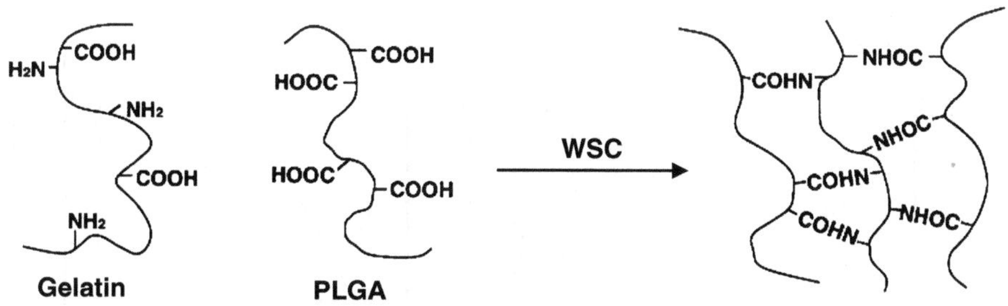

Fig. 4.26 Schematic representation of the gelatin-poly(L-glutamic acid) (PLGA)-water-soluble carbodiimide (WSC) gelation system

seconds through amide formation between the amino groups of gelatin and carboxyl groups of PLGA, as shown in Fig. 4.26. This gel gives higher bonding strength to various soft tissues than that of clinically used fibrin glues [86, 87]. Degradation of the gel did not cause any significant inflammation in an animal model reference. However, the gelatin aqueous solution, before the addition of WSC, spontaneously sets to a physical gel at temperatures around 25°C and below. This problem was overcome by the addition of urea without altering the WSC cross-linking characteristics and tissue adhesion properties [88]. The hemostatic capability of this gel was examined using an injury model with constant blood oozing in dog spleen [89]. Compared with fibrin glue, the gelatin-PLGA gel yielded a significantly higher rate of successful hemostasis due to the strong adhesion of the glue and rapid gelation of the solution mixture. The strong adhesion was thought to be due to the direct reaction of WSC-activated PLGA with amino groups of the tissue. When the effectiveness of the gelatin-PLGA gel was investigated as a lung air-leak sealant, the gel was also found to adhere strongly to the lung surface and to exhibit a superior sealant effect compared to that of fibrin glue [90].

To eliminate the use of WSC, a glue composed of *N*-hydroxysuccinimide (NHS)-activated PLGA and gelatin was developed [91]. The NHS-activated PLGA could be stored for an extended period of time under dry, cool conditions without losing cross-linking ability. Cross-linked gel spontaneously formed when the NHS-activated PLGA was mixed with gelatin, and the bonding strength of this glue with natural tissues was higher than that of fibrin glue. In another study, a gel composed of amine-modified gelatin and aldehyde-modified polysaccharides was synthesized. The amino content of gelatin increased upon reaction of gelatin with ethylene diamine in the presence of WSC [92]. Dextran and hydroxyethyl starch were oxidized by sodium periodate to convert 1,2-hydroxyl groups into dialdehyde groups. When the solutions of gelatin and modified dextran or hydroxyethyl starch were mixed, gel formed under the formation of Schiff base between the amino groups of gelatin and the aldehydes of polysaccharides. As shown in Fig. 4.27, higher bonding strengths (max. 225 $gfcm^{-2}$) could be achieved as well as faster gelation time (2 s) with increasing amine and aldehyde groups.

Sung et al. investigated several cross-linking agents (epoxy compound, WSC, and genipin) for gelatin cross-linking. Their properties as tissue adhesives were compared with those of GRF (gelatin–resorcinol–formaldehyde) and GRG (gelatin–resorcinol–glutaraldehyde) [31]. Genipin is a natural cross-linking agent which is

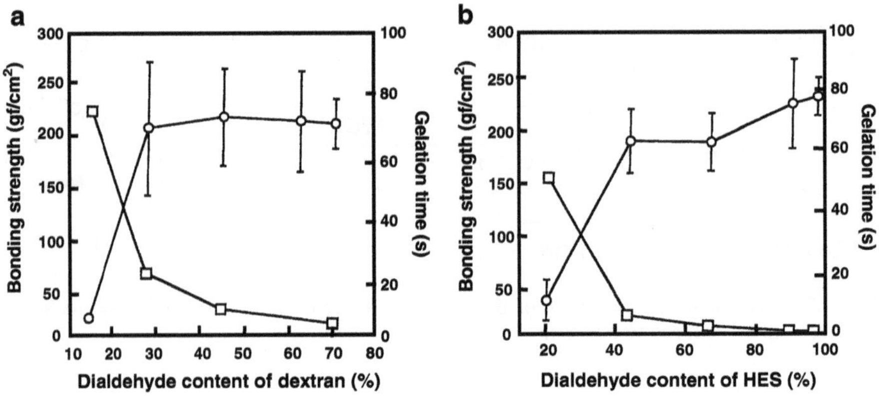

Fig. 4.27 Bonding strength (*circle*) and gelation time (*square*) of gelatin-polysaccharide gel according to the dialdehyde content. (**a**) Dextran and (**b**) hydroxyethyl starch (HES) [92]

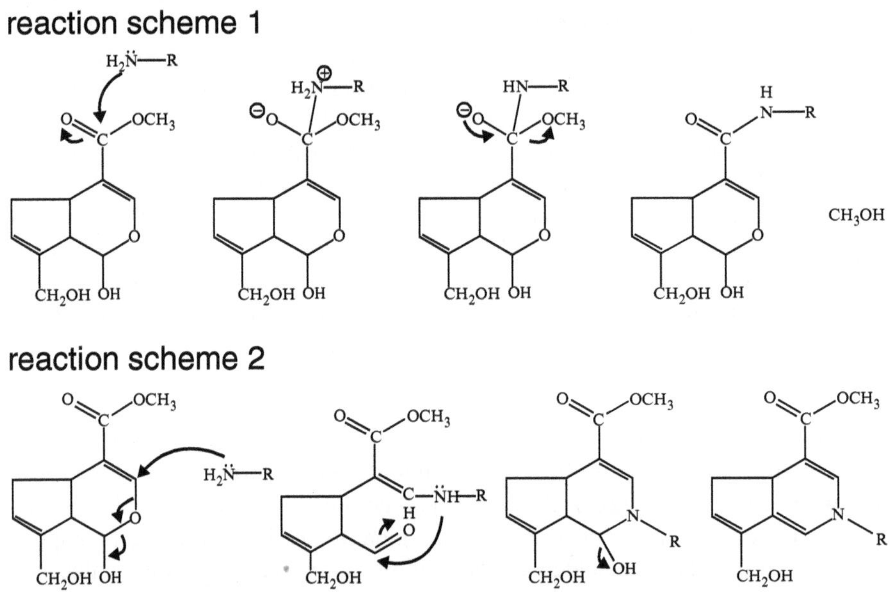

Fig. 4.28 Structure of genipin and its reaction with an amine compound [93]

obtained from geniposide, an iridoid glucoside isolated from the fruits of *Genipa americana* and *Gardenia jasminoides Ellis*. The chemical structure of genipin and the scheme of its reaction with the amine group is shown in Fig. 4.28 [93]. It is believed that two different sites (ester group (in scheme 1) and C3 carbon atom (in scheme 2)) on the genipin molecule were involved in the cross-linking reaction. Cytotoxicity tests using 3T3 fibroblasts revealed that GRF, GRG, and gelatin cross-linked with an epoxy compound (GRE) were highly toxic, whereas gelatins cross-linked with WSC and alginate (GAC) or with genipin (GG) had no toxic effects on 3T3 fibroblast

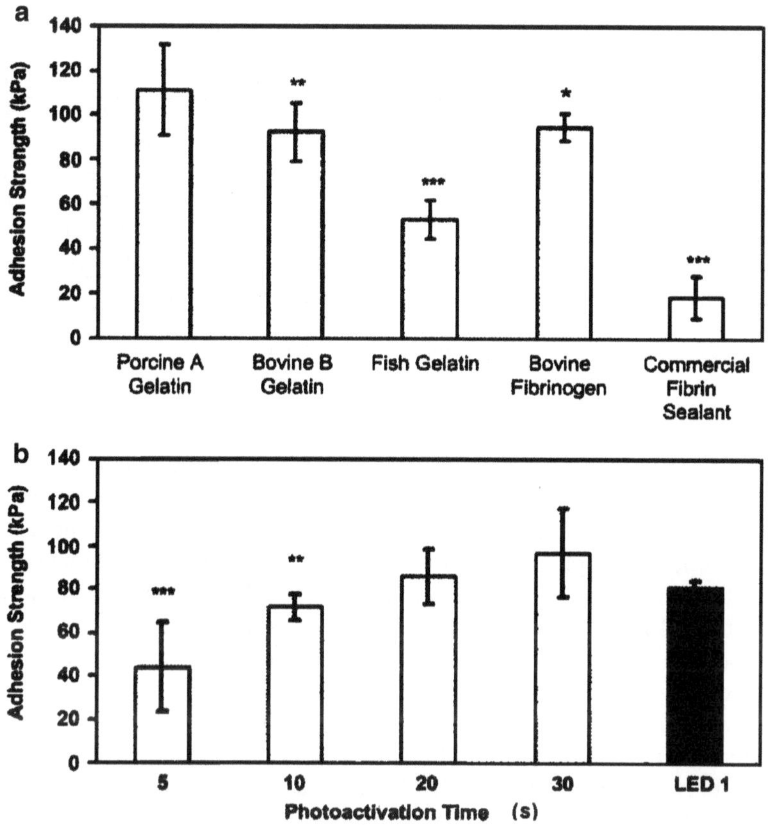

Fig. 4.29 Adhesive strength of cross-linked materials. (**a**) Comparison of photo-polymerized gelatins and bovine fibrinogen by using a xenon lamp (for 30 s) and a commercial tissue sealant (Tisseel). (**b**) The effect of photo-activation time and light source on the adhesion strength of porcine gelatin was determined using a xenon lamp (*white bars*) or a light-emitting diode (LED) lamp (*black bar*) [94]

growth even when a high concentration of cross-linkers (10,000 ppm) was used. GRE exhibited no bonding strength, whereas GAC and GG exhibited strengths of 76.2 ± 11.2 and 69.5 ± 5.6 g/cm^2, respectively. By comparison, GRF and GRG exhibited higher bonding strengths (137.9 ± 25.7 and 152 ± 23.6 g/cm^2, respectively) and yielded stiffer gels than gels cross-linked with other agents. The authors therefore concluded that GAC and GG were preferable when minimal cytotoxicity and stiffness of adhesives were required.

Various gelatins were photo-cross-linked and evaluated [94]. As previously described for photo-cross-linking of fibrinogen, gelatin was also rapidly cross-linked using blue light, a ruthenium catalyst, and a persulfate oxidant. The gelatins used were porcine type A (~300 g Bloom), bovine type B (~225 g Bloom), and cold-water fish gelatin (~60 kDa). The adhesive strengths of these materials are shown in Fig. 4.29. The adhesive strength of porcine gelatin was similar to that of

Table 4.16 Physical properties of cross-linked protein biomaterials [94]

	Bovine fibrinogen (*n*=7)	Bovine gelatin (n=3)	Porcine gelatin	
			U (*n*=4)	BH (*n*=4)
Engineering stress at break (kPa)	46±8	82±9	338±48	91±6
True stress at break (kPa)	75±13	592±97	2,615±455	223±29
Extension to break (%)	62±11	587±46	672±45	145±16
Elastic modulus (kPa)	70±6	6±1	14±2	67±6
Swelling% (4 h)	nd	nd	103±11	5±0.2
Swelling% (24 h)	nd	nd	241±29	10±0.3

U unmodified, *BH* derivatized with Bolton-Hunter reagent, samples contained: bovine fibrinogen, 100 mg/ml; bovine gelatin, 150 mg/ml; porcine gelatin, 175 mg/ml; all samples, 1 mM $[RuII(bpy)_3]^{2+}$ and 20 mM SPS. Elastic modulus was derived from the engineering stress at 100% strain. *nd* not determined

photo-polymerized fibrinogen and approximately five times higher than that of the commercial fibrin glue. The maximum adhesion strength was reached within 30 s, but with an LED lamp, high adhesive strength was achieved instantaneously. The physical properties of photo-cross-linked gels are summarized in Table 4.16. The photo-polymerized porcine gelatin swelled substantially with increase in weight of 240% within 24 h. To reduce swelling, the gelatin was derivatized to increase content of phenolic (Tyr-like) residues using Bolton-Hunter reagent (*N*-succinimidyl-3-[4-hydroxy]propionate) and thus to increase its cross-linking density after polymerization. The resultant gel increased the elastic modulus (stiffness) approximately fivefold and reduced its swelling to less than 10% at 24 h. In an animal study, the photo-polymerized gelatin effectively sealed a wound in lung tissue and was not cytotoxic and did not produce any inflammatory response.

PEG-Based Sealant

PEG/dextran aldehyde was developed as biocompatible tissue adhesive [95]. It consisted of a relatively short-chain PEG polymer (linear PEG of 2 kDa or 8-arm PEG of 10 kDa) containing amine-terminated groups and dextran-containing aldehyde groups. Figure 4.30 represents the reaction between PEG amine and dextran aldehyde. The swelling and onset of degradation are shown in Fig. 4.31. Aldehyde-to-amine ratios of 1, 3, and 10 were examined for linear and star-PEG-based constructs. The adhesive forces of these polymers were determined by the number of aldehyde groups within the gel and reached saturation at a dextran aldehyde content of ~20 wt.%. Cytotoxicity and proliferation of 3T3 rat fibroblast tests revealed that PEG/dextran was significantly less cytotoxic than octyl cyanoacrylate. The higher-molecular-weight dextran aldehyde exhibited lower cytotoxicity as well as higher proliferation [95].

To bind regenerated cartilage tissue with an existing one, Wang and colleagues developed a glue on the basis of functionalized chondroitin sulfate with two organic

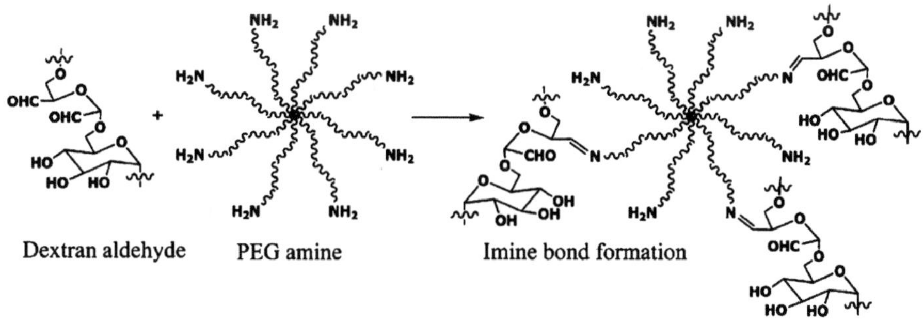

Fig. 4.30 Reaction between dextran aldehyde and poly(ethylene glycol) (PEG) amine [95]

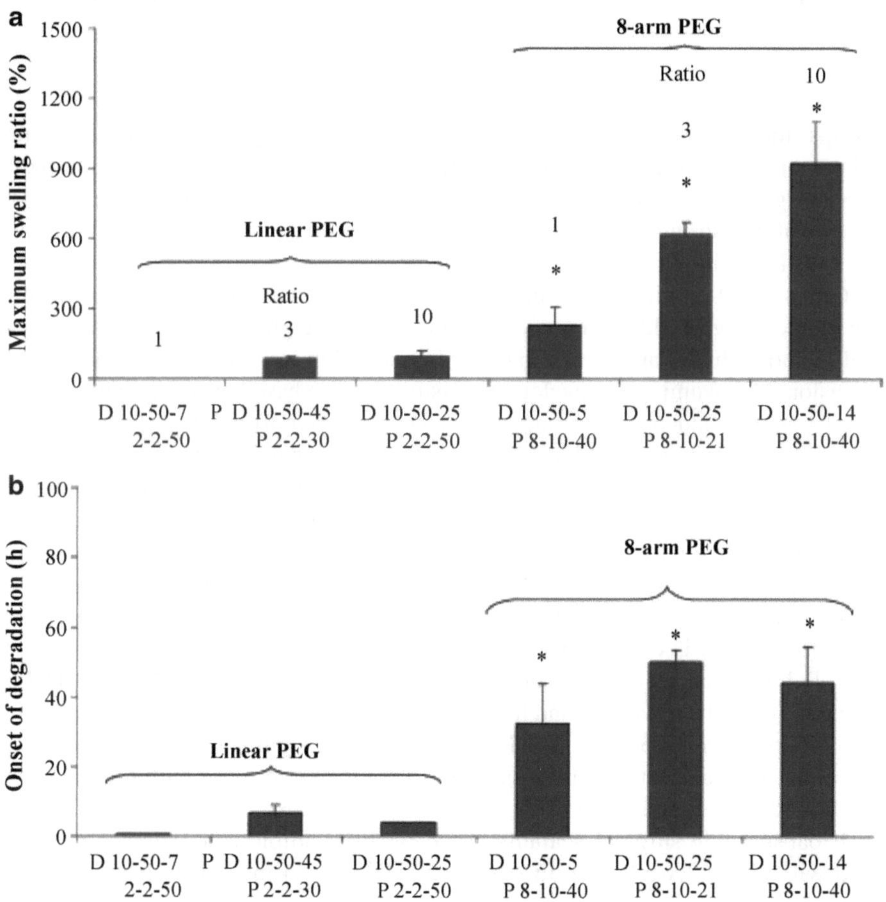

Fig. 4.31 Maximum swelling (**a**) and onset of degradation (**b**) for 2- and 8-arm poly(ethylene glycol) (PEG) [95]

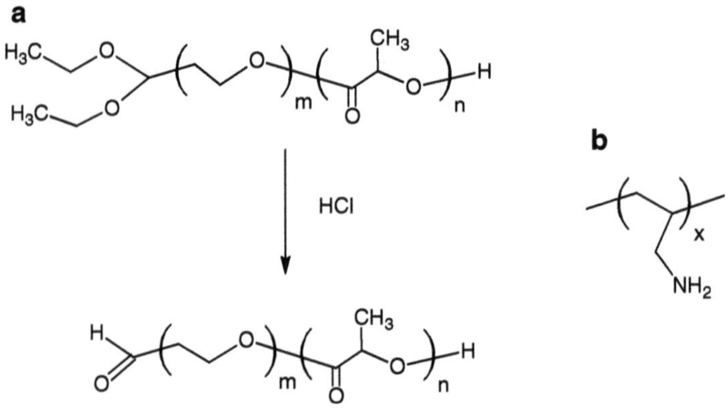

Fig. 4.32 Molecular structures of (**a**) Acetal-terminated poly(ethylene glycol-block-DL-lactide) and (**b**) polyallylamine [97]

groups, methacrylate and aldehyde [96]. The aldehyde groups react with amine groups of collagen in the native tissue to form covalent bonds. In addition, methacrylate groups solidify the implanted material by polymerization reaction. The glue exhibited strong mechanical bonding to the native cartilage and did not injure the cells in the implant or the native cartilage.

Glues based on high-molecular-weight aldehydes were investigated because of the cytotoxicity of aldehydes with low molecular weights (such as formaldehyde and glutaraldehyde) and their capacity of penetration in tissues due to their strong diffusion. Murakami et al. developed a tissue adhesive based on a polymeric micelle consisting of an aldehyde-terminated poly(ethylene glycol)-poly(D,L-lactide) (PEG-PLA) block polymer and polyallylamine [97]. Their chemical structures are shown in Fig. 4.32. Hydrogel rapidly formed (2 s) when polymeric micelles and polyallylamine solutions were mixed in vitro. The hydrogel also cross-linked rapidly in situ under adhesion to mouse peritoneum.

A topical hemostatic agent comprised of polycarbonate of dihydroxyacetone (pDHA) and methoxypoly(ethylene glycol) (MPEG) was studied by Henderson et al. [98]. DHA is the fifth metabolite of glucose as it is metabolized to pyruvic acid. Polymerized sequence of dihydroxyacetone was PEGylated (MPEG-pDHA). The edge of rat liver was cut and treated with MPEG-pDHA (50 mg), normal saline (0.5 ml), or Instat™ (50 mg). The MPEG-pDHA had significantly decreased bleeding time (97 s) and total blood loss (1.35 g) compared to those with normal saline (464 s and 3.83 g) and Instat (165 s and 2.04 g).

Bioabsorbable surgical sealants based on PEG-adipic acid esters functionalized with either 2,4-tolylene diisocyanate (TDI) or 4,4-methylene-bis(phenyl isocyanate) (MDI) were developed, and their biocompatibility and possibility for vascular closure were evaluated [99]. For the synthesis of sealants, di-PEG-adipate was first synthesized by a bulk polycondensation between PEG-600 and adipoyl chloride, and then functionalized with TDI or MDI, followed by reaction with

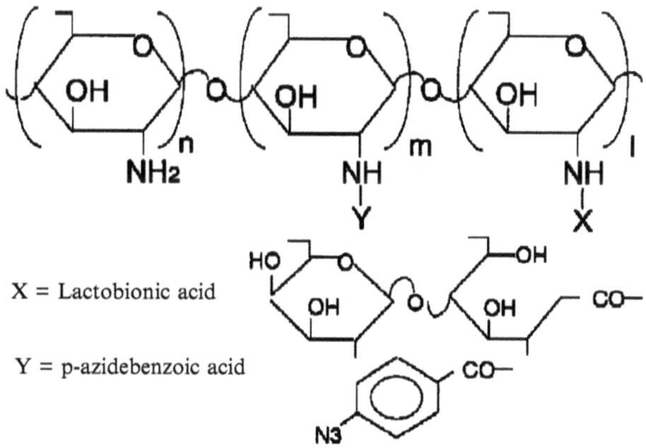

Fig. 4.33 Chemical structure of photo-cross-linkable Az-CH-LA [100]

Table 4.17 Summary of air-sealing strength of chitosan hydrogel and fibrin glue ($n=3$) [100]

	Sealing strength		
	Small intestine mmHg	Aorta mmHg	Trachea mm H_2O
Chitosan gel from 30 mg/ml of Az-CH-LA	61 ± 1	200 ± 18	103 ± 24
Fibrin glue	51 ± 6	80 ± 20	57 ± 20

1,1,1-trimethylolpropane to form partially branched polymers. An abdominal aorta rat arteriotomy closure model was used to evaluate the sealants. It was found that the TDI-functionalized polymer showed sufficient adhesive and cohesive strength and performed better than the MDI-functionalized polymer.

Chitosan-Based Sealant

Chitosan, which is an inexpensive and widely available material, has been investigated for its usefulness as a hemostatic material because it is cationic and antimicrobial, but it may not be effective for severe wounds [100, 101]. Ono et al. investigated photo-cross-linkable chitosan molecules which contain both lactose moieties and photoactive azide groups (Az-CH-LA). Their chemical structure is shown in Fig. 4.33. Its sealant effectiveness was compared with that of fibrin glue [100]. The lactose moieties made chitosan more water-soluble at natural pH, and upon ultraviolet light (UV) irradiation, azide groups photo-cross-linked, and an insoluble hydrogel was produced within 60 s. The air-sealing strength of the gel was found to be higher than that of fibrin glue on small intestine, blood vessels, and trachea (Table 4.17). Skin incisions were made to full thickness on the back of mice, and subsequent application of Az-CH-LA and photo-cross-linking resulted in wound contraction and accelerated wound closure and healing.

Table 4.18 Hemostasis and survival data of TachoComb® (TC) and Az-CH-LA hydrogel combined with a chitosan sponge (PCM-S) [102]

	Control (n=8)	TC (n=8)	PCM-S (n=8)
Nonheparinized rats			
Number of deaths	0	0	0
Bleeding time (min)	6±0.5	1.8±0.2	1.7±0.2
Blood loss (g)	0.5±0.1	<0.3	<0.3
Heparinized rats			
Number of deaths	8	3	0
Bleeding time (min)	28±0.8[a]	25±3[b]	4±1
Blood loss (g)	8.2±0.3[a]	5.0±1.9[b]	1.4±0.5

The data present the mean±SD (n=8). The bars show statistically significance using Sheffé test. *p = 0.000286, **p < 0.0001

[a]Bleeding times (min) and blood losses (g) were measured at the time when the rats died and were measured at the time

[b]When the bleeding completely stopped (n=5) and when the rates died (n=3)

TachoComb® is often used for management of excessive liver hemorrhage following trauma or during selective hepatic resection. However, it does not work well when a patient is heparinized due to inability to form clots. The Az-CH-LA hydrogel was combined with a chitosan sponge (PCM-S), and its hemostatic efficacy was compared with that of TachoComb® in a heparinized rat model [102]. An oozing model was created in rat liver by mechanical incision and penetration with a dermal punch (3 mm diameter). Materials were applied immediately, and amounts of bleeding were measured. The control group had only pressure application to the surface with surgical swabs to stop bleeding. The hemostasis and survival data are shown in Table 4.18. Even under heparinized condition, the bleeding time and blood loss of the PCM-S group were significantly lower than those of the TachoComb® group.

In another study, photo-curable chitosan was evaluated as a new adhesive for peripheral nerve anastomosis to restore continuity to severed peripheral nerves. Low (15 kDa)- and high (50–190 kDa)-molecular-weight chitosans were modified by conjugating with 4-azidobenzoic acid (Az-CH) [103]. Under UV light, Az-CH solution formed a hydrogel in less than 1 min. The low-molecular-weight Az-CH exhibited less swelling and possessed higher rheological storage modulus than the high-molecular-weight one.

Lauto et al. investigated the adhesive ability of chitosan film containing indocyanine green dye (IG) when activated with laser light [104, 105]. As shown in Fig. 4.34,

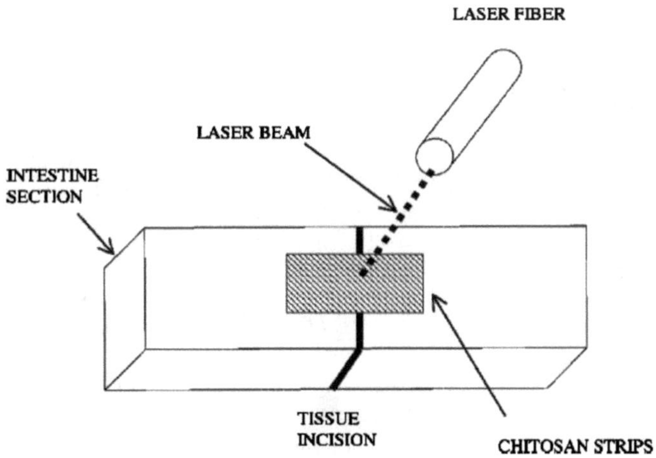

Fig. 4.34 Schematic top view of laser tissue repairing. A strip of chitosan adhesive is applied across the incision, then subsequently irradiated using a laser [105]

a strip of chitosan film was applied across the incision and then subsequently irradiated using a GaAlAs diode laser coupled with a multimode optical fiber to ~15 W/cm^2. The adhesive-bonded sheep intestine with a high tensile strength (~14.7 kPa) demonstrated high E modulus of ~6.8 MPa. The effects of laser power and addition of a natural cross-linker (genipin) to the adhesion composition were investigated for chitosan film impregnated with the antibiotic vancomycin [106]. Laser irradiation did not affect the release or activity of the antibiotic, which survived transient temperatures during infrared irradiation of around 54°C. Adhesive strength also was not affected (15.6 ± 2 kPa).

Photo-chemical tissue bonding (PTB) has been utilized for the adhesion of chitosan film [107]. In the PTB technique, a solution of Rose Bengal is applied between two tissue edges, and the area is irradiated with a green light to cross-link collagen fibers with minimal heat production. The resulting adhesive strength of the film to intestine was 15 ± 2 kPa, and the average temperature of the adhesive increased from 26°C to only 32°C during laser exposure.

Dowling and colleagues developed a chitosan-based amphiphilic biopolymer that exhibits clotting ability in the presence of blood by self-assembly that is readily reversible by introducing a sugar-based supramolecule [108]. First, chitosan was hydrophobically modified by attaching a small number of hydrophobic tails to the backbone (hm-chitosan). The hypothesized mechanism of hemostatic efficacy of the hm-chitosan is shown in Fig. 4.35. The hydrophobes from the chitosan anchor into the hydrophobic interior of blood cell membranes, connecting them into a gel network which can halt the flow of blood. The reaction is not affected by heparinization of blood. To reverse the gel into a sol, α-cyclodextrin, a supramolecular compound with an inner hydrophobic pocket, is added (Fig. 4.36a). The strong affinity of α-cyclodextrin for hydrophobes causes these polymer moieties to unhook from the cell membranes and bind to α-cyclodextrin. As shown in the dynamic rheology

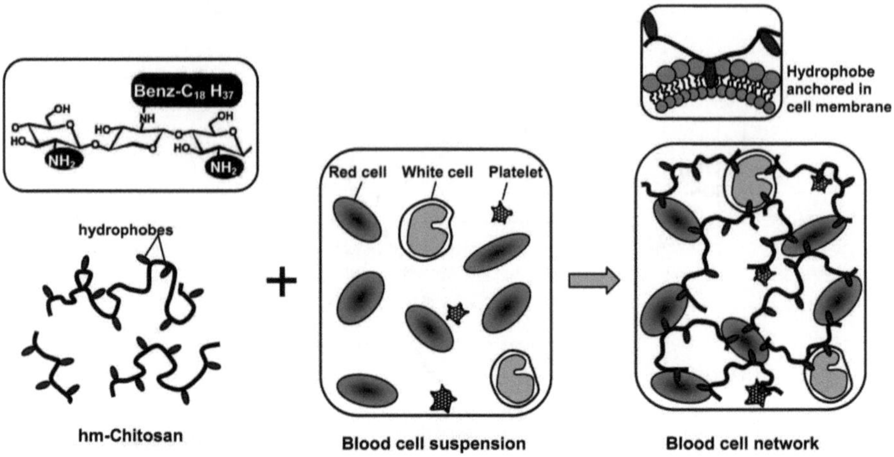

Fig. 4.35 Mechanism of the gelation of blood by hydrophobically modified (hm) chitosan [108]

results in Fig. 4.36b, the initial blood gel based on 0.25 wt.% hm-chitosan exhibits an elastic, gel-like response (closed symbols), whereas addition of 3 wt.% α-cyclodextrin results in a viscous liquid response (open symbols). Preliminary tests with small and large animal injury models demonstrated its increased efficacy at achieving hemostasis (Fig. 4.37).

Polysaccharide-Based Sealant

Adhesives made of aldehyded polysaccharides and ε-poly(L-lysine), which are antibacterial additives for medical use and food, have been investigated for use in sutureless amniotic membrane transplantation [109]. The adhesive gel is prepared using a syringe-like container with two cylinders, one with 2 ml of 14% aldehyded dextran (Mw 75 kDa) solution and the other with 2 mL of 7% ε-poly(L-lysine) (Mw 4 kDa) solution containing 2% acetic anhydride. The acetic anhydride concentration added to the ε-poly(L-lysine) solution determines the gelation time. The gel formed by cross-linking via Schiff base in about 30 s at 37°C degrades within 4 days in vivo. The bonding strength of the glue was four times that of commercial fibrin glue, and almost no cytotoxicity was observed. The efficacy of the glue in fixing amniotic membrane without suturing for ocular surface reconstruction was evaluated in an animal model. Amniotic membrane has been used as surgical material in a wide variety of surgeries of the eye, including ocular surface diseases. It is fixed onto the ocular surface using 10–0 nylon suture. Although this method of suturing facilitates secure attachment, it inflicts trauma on the ocular surface. Suturing also causes postoperative pain and discomfort, associated with complications such as suture abscesses, granuloma formation, and tissue necrosis. The sutureless amniotic membrane placement using the adhesive was safely and successfully performed onto a rabbit sclera surface.

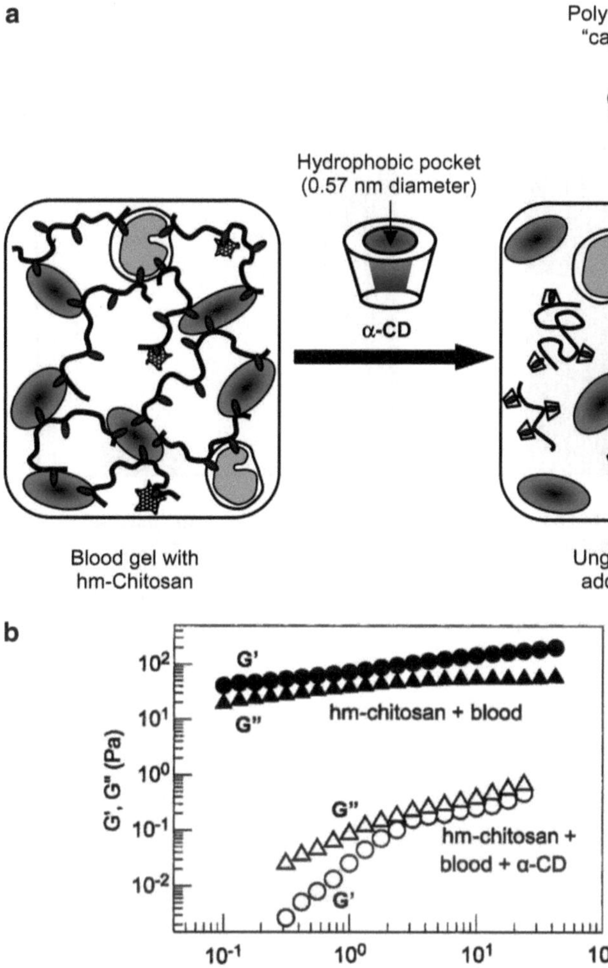

Fig. 4.36 Reversal of blood gelation by α-cyclodextrin (α-CD). (**a**) Reversal mechanism. The α-CD molecule has a barrel shape with a hydrophobic pocket. When added to hydrophobically modified (hm) chitosan/blood gel, the polymer hydrophobes unhook from the cells and become buried within the hydrophobic pockets of the α-CDs. Thus, the connections between the cells are eliminated, and the gel is liquefied. (**b**) The dynamic rheology data that confirm this effect have been shown [108]

Natural Adhesives/Sealants

Mussel-derived adhesive protein is known to be the most powerful natural adhesive and has both flexibility and elasticity [110, 111]. In 1981, Waite and Tanzer discovered 3,4-dihydroxy-L-phenylalanine (DOPA) as a key component for the wet-resistant

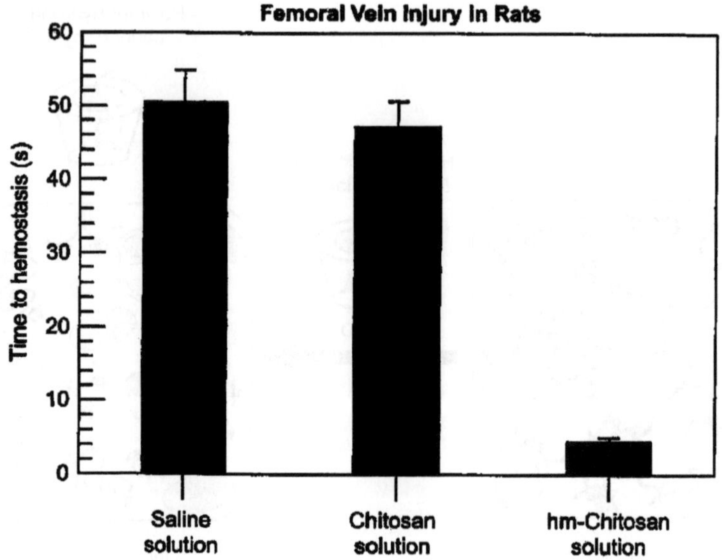

Fig. 4.37 Evaluation of hydrophobically modified (hm) chitosan as a hemostatic agent in a rat model of femoral vein injury [108]

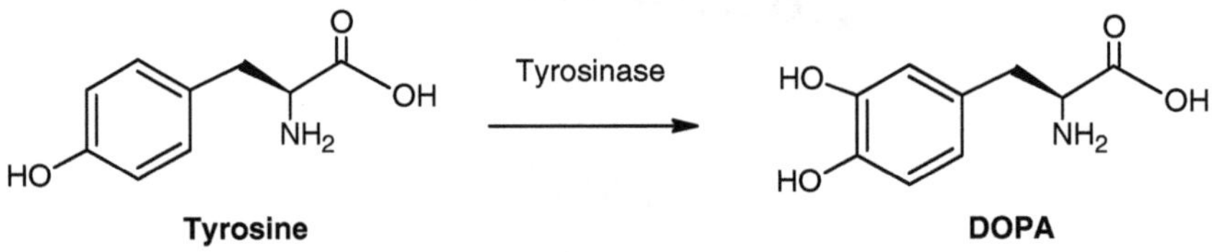

Fig. 4.38 Posttranslational modification of tyrosine residues in mussel adhesive protein

adhesion of mussel adhesive proteins [112]. The chemical structure of DOPA is shown in Fig. 4.38. Nontoxic and nonimmunogenic effects of mussel proteins have been reported [113–115]. More interestingly, it maintains its adhesion in wet environment and adheres to virtually any types of synthetic and natural surfaces [116]. However, expensive extraction, which requires 10,000 mussels to obtain 1 g of one type of adhesive proteins, and unsuccessful large-scale production limit its practical applications [110].

DOPA has been used to increase the adhesion strength of PEG-based sealant [117]. Adhesive consisting of star poly(ethylene glycol) amine and linear dextran aldehyde was developed, but after swelling of the gel, the strength of adhesion to soft tissues significantly decreased. Incorporation of DOPA into the PEG/dextran sealant was investigated to determine enhancement of postswelling sealant performance. Homogeneous solutions of DOPA and lyophilized dextran aldehyde were dialyzed against water followed by freeze-drying. The PEG/dextran doped with 3 mM DOPA aldehyde swelled 50% less, had threefold greater stiffness, and had

50% greater functional adhesive strength than the neat hydrogel. Increasing the DOPA concentration to 11 mM decreased the swelling and mitigated loss of properties with hydration but reduced the initial functional adhesive strength, material modulus, and biocompatibility.

Dermal exudate from *Notaden bennetti* frogs, another natural adhesive, was investigated for medical use [118]. The exudate rapidly forms a protein-based pressure-sensitive adhesive that functions well in wet environments. The glue implanted subcutaneously in mice was bioabsorbed and exhibited no long-term adverse effects. The nonprotein components of the frog glue caused local and transient necrosis, although they did not contribute to the adhesive properties of the glue. The glue could bond severed cartilage tissue both ex vivo and in vivo.

The hemostatic characteristics of keratin derived from human hair were investigated using a rabbit model of lethal liver injury [119]. Keratin was extracted from human hair, and the protein solution was concentrated to 20 wt.% using a rotary evaporation system prior to surgery and exposed to air overnight to form a cross-linked hydrogel. The keratin gel exhibited adhesion to the tissue, and when deposited onto the bleeding surface of liver, it was sufficiently adhesive and absorbed blood and became more adherent within a few minutes. The keratin gel demonstrated efficacy in arresting hemorrhage and improving survivability similar to those of two commercial products, QuickClot and HemCon®. The gel was nontoxic and nonimmunogenic.

Gecko-inspired tissue adhesive, which mimics the nanotopography of gecko feet, was investigated to determine its adhesive properties in a dry environment without chemical glue. The adhesive footpads of geckos are decorated with a dense array of fibrils (setae). Each seta has numerous terminal projections (spatula) that are 200–500 nm in length, and the combination of van der Waals and capillary forces controls the adhesion of these spatulae to the surface. The fibrillar design also enhances interface compliance and conformability to surfaces with a variety of roughnesses. For tissue application, the surface should be optimized to adhere on a wet surface.

Lee et al. investigated the synthetic gecko-adhesive surface, which is effective under water due to reversible noncovalent bonding to inorganic surfaces, by coating the surface with a DOPA mimetic polymer, poly(dopamine methacrylamide-co-methoxyethyl acrylate) (poly(DMA-co-MEA)) [111]. This copolymer was synthesized by free radical polymerization with the adhesive monomer, DMA, which accounts for 17% of this copolymer by weight. As shown in Fig. 4.39, electron beam lithography was used to create an array of holes in a PMMA thin film, and poly(dimethyl siloxane) (PDMS) was cast onto the master and cured to create gecko-foot-mimetic nanopillar arrays. Finally, the surface was coated with poly(DMA-co-MEA). Atomic force microscopy was used to measure the adhesive force of each pillar. The adhesive forces per pillar were 39.8 ± 2 nN (gecko in air), 5.9 ± 0.2 nN (gecko in water), 120 ± 6 nN (gecko + DOPA in air), and 86.3 ± 5 nN (gecko + DOPA in water). Coating with the muscle-mimetic polymer increased its wet adhesive force nearly 15-fold.

The surface of poly(glycerol-co-sebacate acrylate)(PGSA), which is a tough biodegradable elastomer, was modified to mimic the nanotopography of gecko feet [120]. Tissue adhesion was optimized by varying the dimensions of the nanoscale

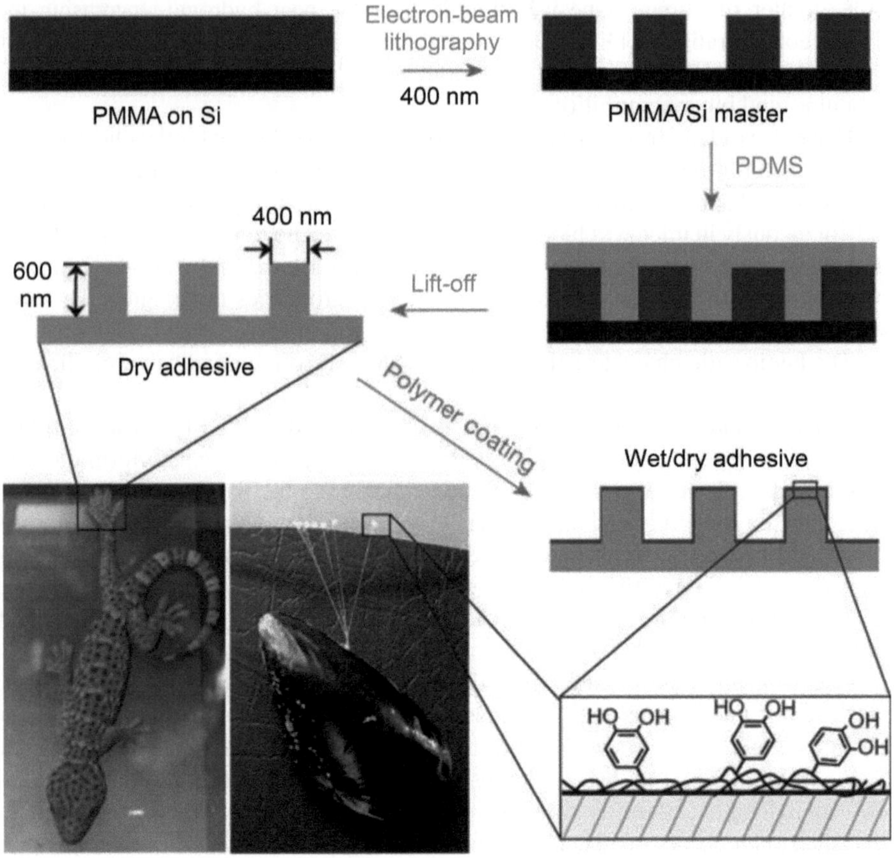

Fig. 4.39 Rational design and fabrication of a wet/dry hybrid nanoadhesive. Electron beam lithography is used to create an array of holes in a poly(methyl methacrylate) (PMMA) thin film supported on Si (PMMA/Si master). Polydimethylsiloxane (PDMS) is casted onto the master, cured, and after lift-off, the gecko-foot-mimetic polymer is coat onto the fabricated nanopillars. The topmost organic layer contains catechols, a key component of wet adhesive proteins found in mussel holdfasts [111]

pillars, including the ratio of tip diameter to pitch and the ratio of tip diameter to base diameter. Coating of these nanomolded pillars of biodegradable elastomers with a thin layer of oxidized dextran with aldehyde groups significantly increased interfacial adhesive strength on porcine intestine tissue in vitro and in the rat abdominal subfascial tissue in an in vivo environment.

Conclusion

Currently, several surgical adhesives, such as fibrin glue, cyanoacrylates, PEG derivatives, and GRF® glue, are clinically available as described above, but have limitations. There is no perfect glue, as the requirements differ depending on the application site and purpose. Although fibrin glue has been most widely used in various applications, its adhesive strength is not sufficient in various surgical situations. In addition, due to the risk of viral infection from natural materials, trends in research on synthetic and more processed polymers with higher adhesion strength than fibrin glue have been emerging. However, it should be noted that nontoxicity, handling, and preparatory time are also important factors in clinical application.

References

1. Matsuda S, Ikada Y (2004) Glues. In: Bowlin GL, Wnek G (eds) Encyclopedia of biomaterials and biomedical engineering. Marcel Dekker, NY
2. Gosain AK, Lyon VB (2002) The current status of tissue glues: Part II. For adhesion of soft tissues. Plast Reconstr Surg 110(6):1581–4
3. Jackson MR (2001) Fibrin sealants in surgical practice: an overview. Am J Surg 182:1S–7S
4. Morikawa T (2001) Tissue sealing. Am J Surg 182:29S–35S
5. Ennker IC, Ennker J, Schoon HD, Rimpler M, Hetzer R (1994) Formaldehyde-free collagen glue in experimental lung gluing. Ann Thorac Surg 57:1622–7
6. Kessler CM, Ortel TL (2009) Recent developments in topical thrombins. Thromb Haemost 101:15–24
7. Burnouf T (2011) Recombinant plasma proteins. Vox Sang 100:68–83
8. Sawamura Y, Asaoka K, Terasaka S, Tada M, Uchida T (1999) Evaluation of application techniques of fibrin sealant to prevent cerebrospinal fluid leakage: a new device for the application of aerosolized fibrin glue. Neurosurgery 44(2):332–7
9. Baker JW, Spotnitz WD, Nolan SP (1987) A technique for spray application of fibrin glue during cardiac operations. Ann Thorac Surg 564–565(43):5
10. Redl H, Schlag G, Dinges HP (1982) Methods of fibrin seal application. Thorac Cardiovasc Surg 30(4):223–7
11. Minato N, Shimokawa T, Katayama Y, Yamada N, Sakaguchi M, Itoh M (2004) New application method of fibrin glue for more effective hemostasis in cardiovascular surgery: rub-and-spray method. Jpn J Thorac Cardiovasc Surg 52(8):361–6
12. Minato N, Katayama Y, Yunoki J, Kawasaki H, Satou H (2009) Hemostatic effectiveness of a new application method for fibrin glue, the "rub-and-spray method", in emergency aortic surgery for acute aortic dissection. Ann Thorac Cardiovasc Surg 15(4):265–71
13. Miyamoto H, Futagawa T, Wang Z, Yamazaki A, Morio A, Sonobe S, Izumi H, Hosoda Y, Hata E (2003) Fibrin glue and bioabsorbable felt patch for intraoperative intractable air leaks. Jpn J Thorac Cardiovasc Surg 51(6):232–6
14. Yano S, Tsuiki H, Kudo M, Kai Y, Morioka M, Takeshima H, Yumoto E, Kuratsu J (2007) Sellar repair with resorbable polyglactin acid sheet and fibrin glue in endoscopic endonasal transsphenoidal surgery. Surg Neurol 67(1):59–64
15. Itano H (2008) The optimal technique for combined application of fibrin sealant and bioabsorbable felt against alveolar air leakage. Eur J Cardiothorac Surg 33(3):457–60
16. Ueda K, Ianaka T, Li TS, Tanaka N, Hamano K (2010) Sutureless pneumostasis using bioabsorbable mesh and glue during major lung resection for cancer: who are the best candidates? J Thorac Cardiovasc Surg 139(3):600–5
17. Sugawara T, Itoh Y, Hirano Y, Higashiyama N, Shimada Y, Kinouchi H, Mizoi K (2005) Novel dural closure technique using polyglactin acid sheet prevents cerebrospinal fluid leakage after spinal surgery. Neurosurgery 57(4 Suppl):290–4
18. Hida K, Yamaguchi S, Seki T, Yano S, Akino M, Terasaka S, Uchida T, Iwasaki Y (2006) Nonsuture dural repair using polyglycolic acid mesh and fibrin glue: clinical application to spinal surgery. Surg Neurol 65(2):136–42

19. Terasaka S, Iwasaki Y, Shinya N, Uchida T (2006) Fibrin glue and polyglycolic Acid nonwoven fabric as a biocompatible dural substitute. Neurosurgery 58(1 Suppl):134–9
20. Shimada Y, Hongo M, Miyakoshi N, Sugawara T, Kasukawa Y, Ando S, Ishikawa Y, Itoi E (2006) Dural substitute with polyglycolic acid mesh and fibrin glue for dural repair: technical note and preliminary results. J Orthop Sci 11(5):454–8
21. Hayashibe A, Sakamoto K, Shinbo M, Makimoto S, Nakamoto T (2006) New method for prevention of bile leakage after hepatic resection. J Surg Oncol 94(1):57–60
22. Joseph T, Adeosun A, Paes T, Bahal V (2004) Randomised controlled trial to evaluate the efficacy of TachoComb H Patches in controlling PTFE suture–hole bleeding. Eur J Vasc Endovasc Surg 27(5):549–52
23. Haas S (2006) The use of a surgical patch coated with human coagulation factors in surgical routine: a multicenter postauthorization surveillance. Clin Appl Thromb Hemost 12(4):445–50
24. Marion Reddy AS, Reddy B, Holzer A, Saringer W, Steiger C, Matula C (2003) Watertightness and effectiveness of a fibrinogen-based collagen fleece (TachoComb®) in neurosurgery. Eur Surg 35(5):278–81
25. Oz MC, Rondinone JF, Shargill NS (2003) Floseal matrix: new generation topical hemostatic sealant. J Card Surg 18:486–93
26. Miyamoto K, Masuda K, Inoue N, Okuma M, Muehleman C, An HS (2006) Anti-adhesion properties of a thrombin-based hemostatic gelatin in a canine laminectomy model: a biomechanical, biochemical, and histologic study. Spine (Phila Pa 1976) 31(4):E91–7
27. Achneck HE, Sileshi B, Jamiolkowski RM, Albala DM, Shapiro ML, Lawson JH (2010) A comprehensive review of topical hemostatic agents: efficacy and recommendations for use. Ann Surg 251(2):217–28
28. Ibrahim MF, Aps C, Young CP (2002) A foreign body reaction to Surgicel mimicking an abscess following cardiac surgery. Eur J Cardiothorac Surg 22(3):489–90
29. Wedmore I, Wedmore I, McManus JG, Pusateri AE, Holcomb JB (2006) A special report on the chitosan-based hemostatic dressing: experience in current combat operations. J Trauma 60(3):655–8
30. Xie H, Khajanchee YS, Teach JS, Shaffer BS (2008) Use of a chitosan-based hemostatic dressing in laparoscopic partial nephrectomy. J Biomed Mater Res B Appl Biomater 85B(1): 267–71
31. Sung H-W, Huang D-M, Chang W-H, Huang R-N, Hsu J-C (1999) Evaluation of gelatin hydrogel crosslinked with various crosslinking agents as bioadhesives: in vitro study. J Biomed Mater Res 46:520–30
32. Laas J, Laas J, Jurmann MJ, Heinemann M, Borst HG (1992) Advances in aortic arch surgery. Ann Thorac Surg 53(2):227–32
33. Albes JM, Krettek C, Hausen B, Rohde R, Haverich A, Borst HG (1993) Biophysical properties of the gelatin-resorcin-formaldehyde/glutaraldehyde adhesive. Ann Thorac Surg 56(4):910–5
34. Guilmet D, Bachet J, Goudot B, Laurian C, Gigou F et al (1979) Use of biological glue in acute aortic dissection. Preliminary clinical results with a new surgical technique. J Thorac Cardiovasc Surg 77:516–21
35. Nienaber CA, Eagle KA (2003) Aortic dissection: new frontiers in diagnosis and management: Part I: from etiology to diagnostic strategies. Circulation 108(5):628–35
36. Hagan PG, Nienaber CA, Isselbacher EM, Bruckman D, Karavite DJ, Russman PL, Evangelista A, Fattori R, Suzuki T, Oh JK, Moore AG, Malouf JF, Pape LA, Gaca C, Sechtem U, Lenferink S, Deutsch HJ, Diedrichs H, Marcos y Robles J, Llovet A, Gilon D, Das SK, Armstrong WF, Deeb GM, Eagle KA (2000) The International Registry of Acute Aortic Dissection (IRAD): new insights into an old disease. JAMA 283(7):897–903
37. Beane Freeman LE, Blair A, Lubin JH, Stewart PA, Hayes RB, Hoover RN, Hauptmann M (2009) Mortality from lymphohematopoietic malignancies among workers in formaldehyde industries: the National Cancer Institute Cohort. J Natl Cancer Inst 101(10):751–61
38. Bonchek LI, Braunwald NS (1967) Experimental evaluation of a cross-linked gelatin adhesive in gastrointestinal surgery. Ann Surg 165(3):420–4

39. Kazui T, Washiyama N, Bashar AH, Terada H, Suzuki K, Yamashita K, Takinami M (2001) Role of biologic glue repair of proximal aortic dissection in the development of early and midterm redissection of the aortic root. Ann Thorac Surg 72:509–14
40. Fukunaga S, Karck M, Harringer W, Cremer J, Rhein C, Haverich A (1999) The use of gelatin-resorcin-formalin glue in acute aortic dissection type A. Eur J Cardiothorac Surg 15:564–70
41. von Oppell UO, Karani Z, Harringer W, Cremer J, Rhein C et al (2002) Dissected aortic sinuses repaired with gelatin-resorcin-formaldehyde (GRF) glue are not stable on follow up. J Heart Valve Dis 11:249–257
42. Suzuki S, Imoto K, Uchida K, Takanashi Y (2006) Aortic root necrosis after surgical treatment using gelatin-resorcinol-formaldehyde (GRF) glue in patients with acute type A aortic dissection. Ann Thorac Cardiovasc Surg 12(5):333–40
43. Bingley JA, Gardner MA, Stafford EG, Mau TK, Pohlner PG, Tam RK, Jalali H, Tesar PJ, O'Brien MF (2000) Late complications of tissue glues in aortic surgery. Ann Thorac Surg 69(6):1764–8
44. Hata H, Takano H, Matsumiya G, Fukushima N, Kawaguchi N, Sawa Y (2007) Late complications of gelatin-resorcin-formalin glue in the repair of acute type A aortic dissection. Ann Thorac Surg 83(5):1621–6
45. Kiunihara T, Shiiya N, Matsuzaki K, Murashita T, Matsui Y (2008) Recommendation for appropriate use of GRF glue in the operation for acute aortic dissection. Ann Thorac Cardiovasc Surg 14:88–95
46. Chao H-H, Torchiana DF (2003) BioGlue: Albumin/Glutaraldehyde sealant in cardiac surgery. J Card Surg 18:500–3
47. Quinn JV (2005) Tissue adhesives in clinical medicine, 2nd edn. BC Decker, ON, USA
48. Azadani AN, Matthews PB, Ge L, Shen Y, Jhun CS, Guy TS, Tseng EE (2009) Mechanical properties of surgical glues used in aortic root replacement. Ann Thorac Surg 87(4):1154–60
49. Zehr K (2007) Use of bovine albumin-glutaraldehyde glue in cardiovascular surgery. Ann Thorac Surg 84:1048–52
50. Coselli JS, Bavaria JE, Fehrenbacher J, Stowe CL, Macheers SK, Gundry SR (2003) Prospective randomized study of a protein-based tissue adhesive used as a hemostatic and structural adjunct in cardiac and vascular anastomotic repair procedures. J Am Coll Surg 197(2):243–52
51. Tansley P, Al-Mulhim F, Lim E, Ladas G, Goldstraw P (2006) A prospective, randomized, controlled trial of the effectiveness of BioGlue in treating alveolar air leaks. J Thorac Cardiovasc Surg 132(1):105–12
52. Fürst W, Banerjee A (2005) Release of glutaraldehyde from an albumin-glutaraldehyde tissue adhesive causes significant in vitro and iv vivo toxicity. Ann Thorac Surg 79:1522–9
53. Klimo P, Khalil A, Slotkin JR, Smith ER, Scott RM, Goumnerova LC (2007) Wound complications associated with the use of bovine serum albumin-glutaraldehyde surgical adhesive in pediatric patients. Operative Neurosurg 60:305–9
54. Coover HJ, Joyner FB, Shearer NH Jr, Wicker TH Jr (1959) Chemistry and performance of cyanoacrylate adhesives. Soc Plast Eng 15:413–7
55. Lauto A, Mawad D, Foster LJR (2008) Review: adhesive biomaterials for tissue reconstruction. J Chem Tech Biotech 83:464–72
56. Singer AJ, Thode HC (2004) A review of the literature on octylcyanoacrylate tissue adhesive. Am J Surg 187:238–48
57. Singer AJ, Perry LC, Allen RL Jr (2008) In vivo study of wound bursting strength and compliance of topical skin adhesives. Acad Emerg Med 15(12):1290–4
58. Leggat PA, Smith DR, Kedjarune U (2007) Surgical applications of cyanoacylate adhesive: a review of toxicity. ANZ J Surg 77(4):209–13
59. Wachter D, Brückel A, Stein M, Oertel MF, Christophis P, Böker DK (2010) 2-Octyl-cyanoacrylate for wound closure in cervical and lumbar spinal surgery. Neurosurg Rev 33(4): 483–9

60. Barnett P, Jarman FC, Goodge J, Silk G, Aickin R (1998) Randomised trial of histoacryl blue tissue adhesive glue versus suturing in the repair of paediatric lacerations. J Paediatr Child Health 34(6):548–50
61. Coulthard P, Worthington H, Esposito M, Elst M, Waes OJ (2004) Tissue adhesives for closure of surgical incisions. Cochrane Database Syst Rev 2004(2):CD004287
62. Göktas N, Karcioglu O, Coskun F, Karaduman S, Menderes A (2002) Comparison of tissue adhesive and suturing in the repair of lacerations in the emergency department. Eur J Emerg Med 9(2):155–8
63. Greenwald BD, Caldwell SH, Hespenheide EE, Patrie JT, Williams J, Binmoeller KF, Woodall L, Haluszka O (2003) N-2-butyl-cyanoacrylate for bleeding gastric varices: a United States pilot study and cost analysis. Am J Gastroenterol 98(9):1982–8
64. Kuo MJ, Yeh HZ, Chen GH, Poon SK, Yang SS, Lien HC, Chang CS (2007) Improvement of tissue-adhesive obliteration of bleeding gastric varices using adjuvant hypertonic glucose injection: a prospective randomized trial. Endoscopy 39(6):487–91
65. Tan PC, Hou MC, Lin HC, Liu TT, Lee FY, Chang FY, Lee SD (2006) A randomized trial of endoscopic treatment of acute gastric variceal hemorrhage: N-butyl-2-cyanoacrylate injection versus band ligation. Hepatology 43(4):690–7
66. Yamaoka T, Tabata Y, Ikada Y (1994) Distribution and tissue uptake of poly(ethylene glycol) with different molecular weights after intravenous administration to mice. J Pharm Sci 83(4):601–6
67. Than KD, Baird CJ, Olivi A (2008) Polyethylene glycol hydrogel dural sealant may reduce incisional cerebrospinal fluid leak after posterior fossa surgery. Neurosurgery 63(1):182–6
68. Thavarajah D, De Lacy P, Hussain R, Redfern RM (2009) Postoperative cervical cord compression induced by hydrogel (DuraSeal): a possible complication. Spine 35(1):E25–6
69. Preul MC, Campbell PK, Bennett SL, Muench TR (2006) A unique dual-function device – A dural sealant with adhesion-prevention properties. Europ Musculoskel Rev 2006:41–4
70. Lee G, Lee CK, Bynevelt M (2010) DuraSeal-hematoma: concealed hematoma causing spinal cord compression. Spine 35(25):1522–4
71. Preul MC, Campbell PK, Garlick DS, Spetzler RF (2010) Application of a new hydrogel dural sealant that reduces epidural adhesion formation: evaluation in a large animal laminectomy model. J Neurosurg Spine 12(4):381–90
72. Glickman M, Gheissari A, Money S, Martin J, Ballard JL, CoSeal Multicenter Vascular Surgery Study Group (2002) A polymeric sealant inhibits anastomotic suture hole bleeding more rapidly than gelfoam/thrombin: results of a randomized controlled trial. Arch Surg 137(3):326–31
73. Campbell PK, Bennett SL, Driscoll A, Sawhney AS (2005) Evaluation of absorbable surgical sealants: in vitro testing. covidien. Available from: http://www.surgipeer.com/duraseal_ous/publications.php.
74. Torchiana DF (2003) Polyethylene glycol based synthetic sealants: potential uses in cardiac surgery. J Card Surg 18(6):504–6
75. Ranger WR, Halpin D, Sawhney AS, Lyman M, LoCicero J (1997) Pneumostasis of experimental air leaks with a new photopolymerized synthetic tissue sealant. Am Surgeon 63:788–95
76. Tanaka K, Takamoto S, Ohtsuka T, Kotsuka Y, Kawauchi M (1999) Application of AdvaSeal for acute aortic dissection: experimental study. Ann Thorac Surg 68:1308–13
77. Thompson I. TissuePatch™3 and TissuePatchDural™: a review of constituent materials Adhesive sealant biomaterials [serial on the Internet]. 2009
78. Mandley D. Evaluation of a novel adhesive film for sealing air leaks after lung surgery. Adhesive sealant biomaterials [serial on the Internet]. 2009
79. Kobayashi H, Sekine T, Nakamura T, Shimizu Y (2001) In vivo evaluation of a new sealant material on a rat lung air leak model. J Biomed Mater Res 58(6):658–65
80. Iwasashi M, Sakane M, Saito H, Taguchi T, Tateishi T, Ochiai N (2009) In vivo evaluation of bonding ability and biocompatibility of a novel biodegradable glue consisting of tartaric acid derivative and human serum albumin. J Biomed Mater Res A 90(2):543–8
81. Elvin CM, Danon SJ, Brownlee AG, White JF, Hickey M, Liyou NE, Edwards GA, Ramshaw JA, Werkmeister JA (2010) Evaluation of photo-crosslinked fibrinogen as a rapid and strong tissue adhesive. J Biomed Mater Res A 93(2):687–95

82. Fancy DA, Kodadek T (1999) Chemistry for the analysis of protein–protein interactions: rapid and efficient cross-linking triggered by long wavelength light. Proc Natl Acad Sci USA 96(11):6020–4
83. Suzuki S, Ikada Y (in press) Sealing effects of cross-linked gelatin. J Biomater Appl
84. Suzuki S, and Ikada, Y. Unpublished Work
85. LeMaire SA, Carter SA, Won T, Wang X, Conklin LD, Coselli JS (2005) The threat of adhesive embolization: BioGlue leaks through needle holes in aortic tissue and prosthetic grafts. Ann Thorac Surg 80(1):106–10
86. Otani Y, Tabata Y, Ikada Y (1996) A new biological glue from gelatin and poly(L-glutamic acid). J Biomed Mater Res 31:157–66
87. Otani Y, Tabata Y, Ikada Y (1996) Adhesion of soft tissues by gelatin-polyanion hydrogels. J Adhesion 59:197–205
88. Otani Y, Tabata Y, Ikada Y (1998) Effect of additives on gelatin and tissue adhesion of gelatin-poly(L-glutamic acid) mixture. Biomaterials 19:2167–73
89. Otani Y, Tabata Y, Ikada Y (1998) Hemostatic capability of rapidly curable glues from gelatin, poly(L-glutamic acid), and carbodiimide. Biomaterials 19:2091–8
90. Otani Y, Tabata Y, Ikada Y (1999) Sealing effect of rapidly curable geralin-poly(L-glutamic acid) hydrogel glue on lung air leak. Ann Thorac Surg 67:922–6
91. Iwata H, Matsuda S, Mitsuhashi K, Itoh E, Ikada Y (1998) Biomaterials 19:1869
92. Mo X, Iwata H, Matsuda S, Ikada Y (2000) Soft tissue adhesive composed of modified gelatin and polysaccharides. J Biomater Sci Polym Edn 11(4):341–51
93. Butler MF, Ng Y-F, Pudney PDA (2003) Mechanism and kinetics of the crosslinking reaction between biopolymers containing primary amine groups and genipin. J Polym Sci Part A: Polym Chem 41(24):3941–53
94. Elvin CM, Vuocolo T, Brownlee AG, Sando L, Huson MG, Liyou NE, Stockwell PR, Lyons RE, Kim M, Edwards GA, Johnson G, McFarland GA, Ramshaw JA, Werkmeister JA (2010) A highly elastic tissue sealant based on photopolymerised gelatin. Biomaterials 31(32):8323–31
95. Artzi N, Shazly T, Crespo C, Ramos AB, Chenault HK, Edelman ER (2009) Characterization of star adhesive sealants based on PEG/dextran hydrogels. Macromol Biosci 11(9):754–65
96. Wang D-A, Varghese S, Sharma B, Strehin I, Fermanian S, Gorham J, Fairbrother DH, Cascio B, Elisseeff JH (2007) Multifunctional chondroitin sulphate for cartilage tissue–biomaterial integration. Nature Mater 6:385–92
97. Murakami Y, Yokoyama M, Okano T, Nishida H, Tomizawa Y, Endo M, Kurosawa H (2007) A novel synthetic tissue-adhesive hydrogel using a crosslinkable polymeric micelle. J Biomed Mater Res A 80(2):421–7
98. Henderson PW, Kadouch DJ, Singh SP, Zawaneh PN, Weiser J, Yazdi S, Weinstein A, Krotscheck U, Wechsler B, Putnam D, Spector JA (2010) A rapidly resorbable hemostatic biomaterial based on dihydroxyacetone. J Biomed Mater Res A 93(2):776–82
99. Hadba AR, Belcheva N, Jones F, Abuzaina F, Calabrese A, Kapiamba M, Skalla W, Taylor JL, Rodeheaver G, Kennedy J (2011) Isocyanate-functional adhesives for biomedical applications. Biocompatibility and feasibility study for vascular closure applications. J Biomed Mater Res B Appl Biomater 99B:27–35
100. Ono K, Saito Y, Yura H, Ishikawa K, Kurita A, Akaike T, Ishihara M (2000) Photocrosslinkable chitosan as a biological adhesive. J Biomed Mater Res 49(2):289–95
101. Ishihara M, Nakanishi K, Ono K, Sato M, Kikuchi M, Saito Y, Yura H, Matsui T, Hattori H, Uenoyama M, Kurita A (2002) Photocrosslinkable chitosan as a dressing for wound occlusion and accelerator in healing process. Biomaterials 23(3):833–40
102. Horio T, Ishihara M, Fujita M, Kishimoto S, Kanatani Y, Ishizuka T, Nogami Y, Nakamura S, Tanaka Y, Morimoto Y, Maehara T (2010) Effect of photocrosslinkable chitosan hydrogel and its sponges to stop bleeding in a rat liver injury model. Artif Organs 34(4):342–7
103. Rickett TA, Amoozgar Z, Tuchek CA, Park J, Yeo Y, Shi R (2011) Rapidly photo-cross-linkable chitosan hydrogel for peripheral neurosurgeries. Biomacromolecules 12(1):57–65
104. Lauto A, Hook J, Doran M, Camacho F, Poole-Warren LA, Avolio A, Foster LJ (2005) Chitosan adhesive for laser tissue repair: in vitro characterization. Lasers Surg Med 36(3):193–201

105. Lauto A, Stoodley M, Marcel H, Avolio A, Sarris M, McKenzie G, Sampson DD, Foster LJ (2007) In vitro and in vivo tissue repair with laser-activated chitosan adhesive. Lasers Surg Med 39(1):19–27
106. Foster LJ, Thomson K, Marçal H, Butt J, Watson SL, Wakefield D (2010) Chitosan-vancomysin composite biomaterial as a laser activated surgical adhesive with regional anti-microbial activity. Biomacromolecules 11(12):3563–70
107. Lauto A, Mawad D, Barton M, Gupta A, Piller SC, Hook J (2010) Photochemical tissue bonding with chitosan adhesive films. Biomed Eng 9:47
108. Dowling MB, Kumar R, Keibler MA, Hess JR, Bochicchio GV, Raghavan SR (2011) A self-assembling hydrophobically modified chitosan capable of reversible hemostatic action. Biomaterials 32(13):3351–7
109. Takaoka M, Nakamura T, Sugai H, Bentley AJ, Nakajima N, Fullwood NJ, Yokoi N, Hyon SH, Kinoshita S (2008) Sutureless amniotic membrane transplantation for ocular surface reconstruction with a chemically defined bioadhesive. Biomaterials 29(19):2923–31
110. Cha HJ, Hwang DS, Lim S (2008) Development of bioadhesives from marine mussels. Biotechnol J 3:631–5
111. Lee H, Lee BP, Messersmith PB (2007) A reversible wet/dry adhesive inspired by mussels and geckos. Nature 448:338–41
112. Waite JH, Tanzer ML (1981) Polyphenolic substance of *Mytilus edulis*: novel adhesive containing L-Dopa and hydroxyproline. Science 212(4498):1038–40
113. Dove J, Sheridan P (1986) Adhesive protein from mussles: possibilities for dentistry, medicine and industry. J Am Dent Assoc 112:879
114. Grande DA, Pitman MI (1988) The use of adhesives in chondrocyte transplantation surgery: preliminary studies. Bull Hosp Jt Dis Otrhop Inst 48:140–8
115. Saez C, Pardo J, Gutierrez E, Brito M et al (1991) Immunological studies of the polyphenolic proteins of mussels. Comp Biochem Physiol 98B:569–72
116. Lee H, Dellatore SM, Miller WM, Messersmith PB (2007) Mussel-inspired surface chemistry for multifunctional coatings. Science 318:426–30
117. Shazly TM, Baker AB, Naber JR, Bon A, Van Vliet KJ, Edelman ER (95) Augmentation of postswelling surgical sealant potential of adhesive hydrogels. J Biomed Mater Res A 95(4): 1159–69
118. Graham LD, Danon SJ, Johnson G, Braybrook C, Hart NK, Varley RJ, Evans MD, McFarland GA, Tyler MJ, Werkmeister JA, Ramshaw JA (2010) Biocompatibility and modification of the protein-based adhesive secreted by the Australian frog Notaden bennetti. J Biomed Mater Res A 93(2):429–41
119. Aboushwareb T, Eberli D, Ward C, Broda C, Holcomb J, Atala A, Van Dyke M (2009) A keratin biomaterial gel hemostat derived from human hair: evaluation in a rabbit model of lethal liver injury. J Biomed Mater Res B Appl Biomater 90(1):45–54
120. Mahdavi A, Ferreira L, Sundback C, Nichol JW, Chan EP, Carter DJ, Bettinger CJ, Patanavanich S, Chignozha L, Ben-Joseph E, Galakatos A, Pryor H, Pomerantseva I, Masiakos PT, Faquin W, Zumbuehl A, Hong S, Borenstein J, Vacanti J, Langer R, Karp JM (2008) A biodegradable and biocompatible gecko-inspired tissue adhesive. Proc Natl Acad Sci USA 105(7):2307–12

Chapter 5
Barriers to Prevent Tissue Adhesion

Abstract Adhesion often occurs after surgery, especially in abdominal, gynecological, and thoracic surgeries. It can cause serious postsurgical complications such as infertility, chronic pelvic pain, and difficulties of extensive adhesiolysis in future surgery. Antiadhesive barriers have been developed to provide physical separation between the injured site and the adjacent tissues. Since physical separation is only required during the critical time of wound healing of the damaged area, bioabsorbable polymers have been found to be appropriate. A range of adhesion barriers are commercially available in the forms of solutions, gels, and sheets. Their characteristics and limitations are discussed in this chapter. Recent investigations are also described on newly developed devices that show promising performance such as ease of handling and more effective reduction of adhesion.

Introduction

Postoperative adhesions of internal organs to each other by formation of fibrous tissue have often been observed during cardiac, thoracic, abdominal, pelvic, and gynecologic surgeries such as cesarean sections, colectomies, and hernia repairs. It is thought that postoperative adhesions occur in 93–100% of patients who undergo transperitoneal surgery. Tissue adhesion can cause serious postsurgical complications, such as small-bowel obstruction (SBO), infertility, chronic abdominal pain, and difficulty of second operation. Postoperative adhesion has been reported to be responsible for 60–70% of cases of SBO and 15–40% of infertilities due to prevention of the ovum pick-up mechanism and gamete transport.

Approximately 300,000 hospital readmissions annually are either directly or indirectly related to the formation of postoperative adhesions in the USA, and the annual financial expenditure for pertinent treatment exceeds $1.3 billion [1]. Significant financial costs, the immeasurable cost of patient discomfort, and the need to either eliminate or reduce the incidence and severity of adhesions are all

S. Suzuki and Y. Ikada, *Biomaterials for Surgical Operation*,
DOI 10.1007/978-1-61779-570-1_5,

clear. A wide variety of commercial products are thus available, and the American market for adhesion prevention materials is estimated to be $500–600 million. This chapter introduces commercial products as well as studies of the prevention of tissue adhesions.

Mechanism of Adhesion

The main cause of postoperative adhesion formation is surgical trauma to the peritoneum [2]. Fibrin deposition occurs due to bleeding when trauma occurs on the peritoneal surface in surgery. If the mesothelial surface is present, lysis of existing fibrin deposits by the synthesis of tissue-type plasminogen activator (tPA) occurs as a normal physiological consequence. Tumor necrosis factor (TNF), a proinflammatory cytokine, is released from macrophages at the time of tissue injury and peaks in 1–2 h after injury [3] (Fig. 5.1). The increase in TNF induces the rapid synthesis and release of type 1 plasminogen activator inhibitor (PAI-1) by mesothelial and inflammatory cells. This results in the loss of tPA activity because PAI-1 is the fastest and principal inhibitor of tPA and facilitates adhesion by inhibiting fibrinolysis. A study of peritoneal fluid after surgery has demonstrated that this loss occurs within 6–12 h postoperatively. As illustrated in Fig. 5.2, once the fibrin is deposited, the organs are brought in close proximity compared to normal condition due to swelling, and the fibrin deposits form bridges between the opposing tissue surfaces. As a result of diminished fibrinolytic activity, fibrin deposits are transformed to adhesions.

After peritoneal tissue is traumatized, fibrin deposition and matrix formation are evident within 12 h, and new methothelium normally forms within 7–9 days. The barrier approach to adhesion prevention involves the application of materials to form a temporary mechanical barrier during the critical period of postinflammatory fibrin deposition.

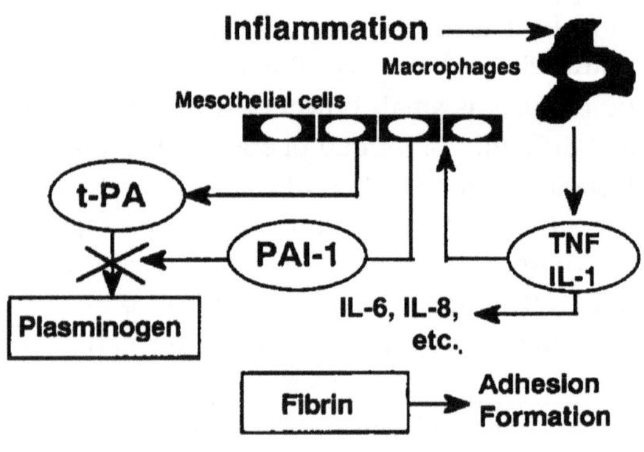

Fig. 5.1 Schematic illustration of the local mechanism of adhesion prevention [3]

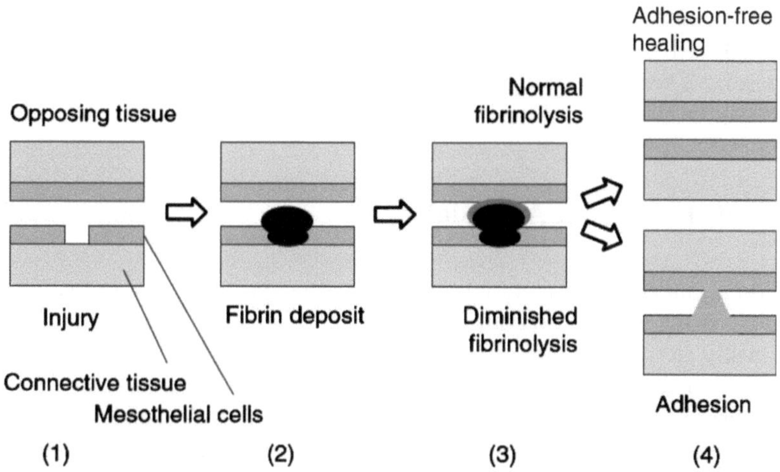

Fig. 5.2 Schematic illustration of the process of normal tissue repair and adhesion formation. (1) On peritoneal injury, (2) a fibrin deposit is formed on the damaged mesothelium, and it enlarges to form a bridge between 2 opposing tissue surfaces. (3) Locally produced fibrinolytic factors are released, and they degrade the entire or part of this fibrin bridge. (4) When the peritoneum is slightly damaged and most of the mesothelial cells are intact, there is a dynamic balance between fibrinolysis and fibrinogenesis, and then, adhesion-free healing takes place. Diminished fibrinolytic activity transforms the fibrin deposit into an adhesion [2]

Preventing Adhesion

Strategies of prevention of adhesions can be grouped into four categories: general principles, surgical techniques, chemical agents, and mechanical barriers [4]. General rules of treatment include avoiding unnecessary peritoneal dissection, avoiding spillage of intestinal contents or gallstones, and the use of starch-free gloves. Surgical techniques include laparoscopic rather than open surgery. Laparoscopy involves minimal handling of tissue, little manipulation of internal organs, and maintenance of tissue moistness due to the closed environment. Nonclosure of the peritoneum has been found to be associated with decreased adhesion formation [5, 6]. In a study, the rate of adhesion formation after laparotomy with peritoneal closure was found to be 22%, compared with 16% without closure [7]. However, other studies found no differences [8, 9]. Chemical agents including fibrinolytic agents, anticoagulants, anti-inflammatory agents, and antibiotics have been used to reduce adhesion formation [4], but their effects are also unclear.

Antiadhesive materials have been developed to prevent tissue adhesion by providing a physical barrier between an injured site and the adjacent tissues. Although several nonabsorbable synthetic materials such as silicone and PTFE have been shown to be effective, bioabsorbable materials are preferred because of the lack of necessity of secondary surgery to remove nonabsorbable materials and the lack of need to consider long-term biocompatibility such as encapsulation of the material, which will also evoke tissue adhesion. Moreover, nonabsorbable synthetic materials have been shown to form adhesions on long-term application [10].

Materials to Prevent Adhesions

An optimal antiadhesive material should possess the following properties (1) it should cover the wound site until it is no longer susceptible to adhesion; (2) disappear quickly after the required time to avoid foreign-body reactions to the material, which can cause other sites of adhesion; and (3) have mechanical properties sufficient for ease of handling. The extent of adhesion formation depends on the patient, as well as the magnitude and type of surgery performed. According to Rodeheaver et al. [11], at least 36 h of antiadhesive application were required to effectively reduce adhesions in healing of injured abdominal surfaces of rat using a silicone film as adhesive barrier. This is well within the range of depressed plasminogen activator activity level [12]. Matsuda et al. also found that 2-day application of the antiadhesive material, poly(vinyl alcohol) film, reduced the incidence rate of adhesions from 90 to 18% in rats [13].

To reduce adhesion formation, a wide variety of substances and materials have been used over the years. Commercially available antiadhesive products are summarized in Table 5.1 and will be discussed below.

Table 5.1 Commercial products used for preventing adhesion

Structure	Name	Composition	Manufacturer
Solution	Hyskon	32% dextran 70	Medisan Pharmaceuticals
	Adept	4% icodextrin solution	Baxter Healthcare
	Intergel	0.5% ferric hyaluronate	Lifecore Biomedical
Gel	FlowGel	Poloxamer 407	Angiotech Pharmaceuticals
	A-Part Gel	PVA and CMC	Aesculap AG and Co
	Sepracoat	HA and CMC	Genzyme Corporation
	Intercoat	PEO and CMC	Ethicon
	SprayGel	PEG	Covidien
	SprayShield	PEG	Covidien
	Adhibit	PEG	Angiotech Pharmaceuticals
Sheet	Preclude	ePTFE	Gore-Tex
	Seprafilm	HA and CMC	Genzyme Corporation
	Interceed	Oxidized cellulose	Johnson and Johnson Ltd.
	CardioWrap and SurgiWrap	Polylactide	MAST Biosurgery
	REPEL-CV	PLA and PEG	SyntheMed, Inc.
	Cova CARD and Cova ABDO	Porcine type I collagen	Biom'Up

PVA polyvinyl alcohol, *CMC* carboxymethyl cellulose, *HA* hyaluronic acid, *PEO* polyethylene oxide, *PEG* poly(ethylene glycol), *ePTFE* expanded polytetrafluoroethylene, *PLA* poly(lactic acid)

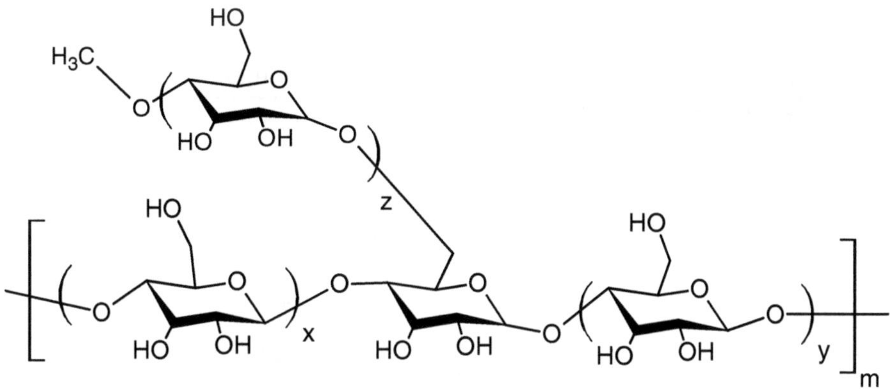

Fig. 5.3 Chemical structure of Adept® (icodextrin)

Solutions

Crystalloid solutions, such as lactated Ringer's solution and plain saline, have been commonly used to prevent adhesions in the abdominal cavity after surgery. In addition to its mechanical effect of separation of raw peritoneal surfaces, it dilutes fibrin and fibrinous exudates released from injured tissues [4]. These fluids are absorbed from the peritoneal cavity at an estimated rate of 35 mL/h, and at least 5 L are needed to cover the first 6 postoperative days. However, such a large amount will increase the risk of infection, fluid overload with pulmonary edema, and leakage at puncture sites. In an attempt to prolong the period of instillate persistence inside the peritoneal cavity, more viscous solutions have been investigated. Hyskon™ (Medisan Pharmaceuticals Inc., Piscataway, NJ) is a solution of 32% dextran 70 and available in the USA. Hyskon™ is left in the abdominal cavity to prevent sticking by causing tissue to literally slide around freely. It is absorbed in 5–7 days. Clinical studies of use of Hyskon™ have revealed mixed results. Although use of Hyskon™ exhibited fewer and less severe adhesions than lactated Ringer's solution, another study reported no difference between these groups [14]. Side effects have also been reported, such as vulvar edema, leg edema, pleural effusion, coagulopathy, and rarely allergic response [15–20]. It is hardly used today.

Adept® (Baxter Healthcare, Deerfield, IL) is a 4% solution of icodextrin, an α-1,4-linked glucose polymer produced by hydrolysis of cornstarch (Fig. 5.3). It was approved for gynecologic laparoscopic surgery by the USFDA in 2006. It is used as an irrigant fluid throughout surgery, and at the end of surgery, 1 L is instilled and left in the peritoneal cavity. A randomized study of laparoscopic gynecologic surgery reported that instillation of Adept® decreased adhesion formation and reformation compared to lactated Ringer's solution instillation [21].

Intergel (Lifecore Biomedical, Inc., Chaska, MN) contains 0.5% cross-linked hyaluronic acid (HA) with ferric ion (Fe) and was approved by FDA in 2001. The reaction scheme and the ideal structure of the Fe-HA network are shown in Fig. 5.4.

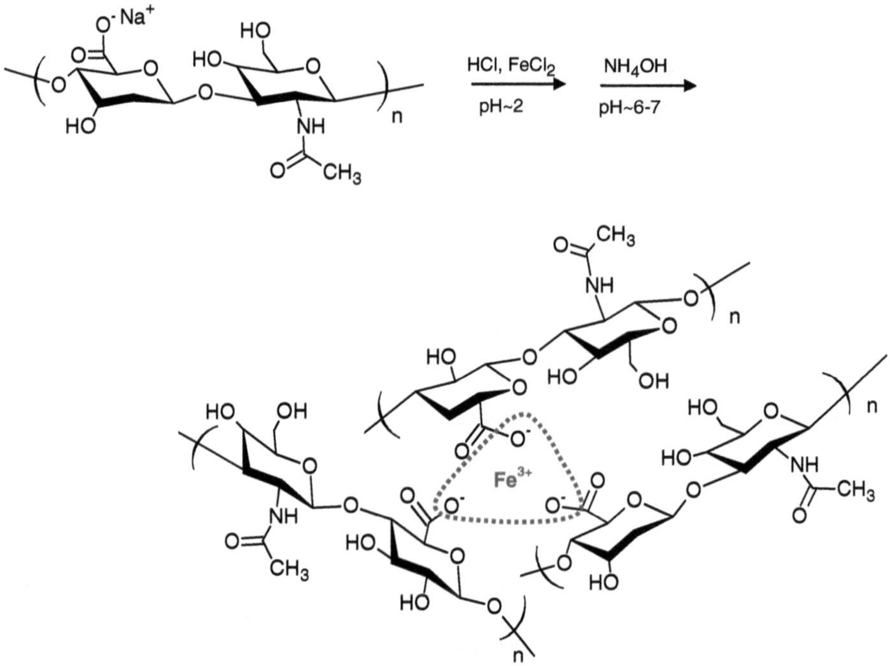

Fig. 5.4 Reaction scheme for the formation of and the idealized structure of the iron–cross-linked hyaluronic acid (Fe-HA) hydrogel

It reduced the extent of adhesion formation following abdominal surgery. In a randomized clinical trial, patients underwent gynecologic surgery and received 300 mL of Intergel solution ($n = 143$) or lactated Ringer's solution ($n = 138$) as an intraperitoneal instillate at the completion of surgery [22]. Adhesion formation was evaluated at second-look laparoscopy 6–12 weeks later. Patients treated with Intergel solution had significantly fewer adhesions compared to the control group. The extent and severity of adhesions were also significantly reduced. However, a high rate of postoperative complications, including postoperative infection and even death, was reported following the use of Intergel, and the manufacturer voluntarily recalled Intergel in 2003 after receiving a warning from FDA [23].

Gels

Gels have been developed for site-specific adhesion prevention devices that can be easily delivered during laparoscopy since using sheets and films via laparoscopy is challenging. FlowGel (Angiotech Pharmaceuticals Inc., Vancouver, BC) consists of Poloxamer 407, which is a surfactant with the ability for reversible conversion from

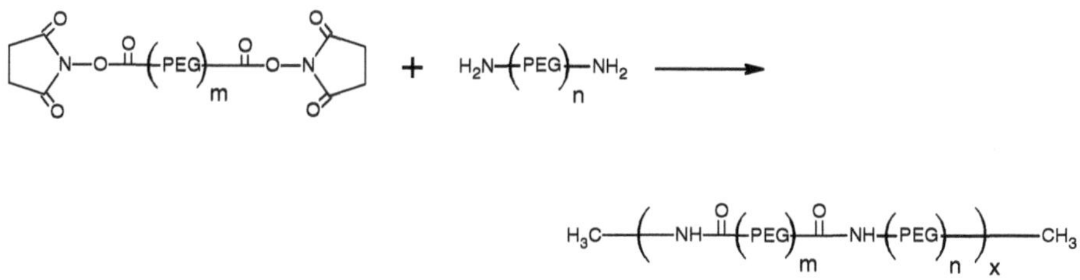

Fig. 5.5 Reaction scheme for the formation of SprayGel

liquid state at room temperature to a gel form at body temperature [24]. In an animal study, Poloxamer 407 was used to cover the uterine and sidewall defects in a rabbit model. Significant reduction of adhesion formation was observed for the treated sides (60% free from adhesion) compared to the untreated control (5%).

A-Part® gel is a bioabsorbable transparent gel composed of PVA and CMC [25]. Material characteristics such as viscosity and adherence to the wound were tested with good results [26, 27]. Preclinical data in animal studies revealed high efficacy in adhesion prevention, and wound and anastomosis healing were not negatively affected by the application of this gel in animal models. No clinical study of this product appears to have been published.

Sepracoat (Genzyme Corp.) is a modification of Seprafilm® which will be discussed later. It consists of hyaluronic acid and CMC in a highly viscous gel form and absorbed in 7 days. A randomized, controlled trial revealed it to be effective, but it did not receive FDA approval, and was withdrawn from the market in 1997.

Intercoat® (Ethicon, Somerville, NJ), a viscoelastic gel, is based on poly(ethylene oxide) and CMC, which is stabilized with calcium. In a pilot study, 28 patients with pelvic adhesions, tubal occlusion, endometriosis, and/or dermoids were randomized to receive Intercoat® or no treatment after surgery [28]. Second-look surgery was performed 6–10 weeks later. Thirty-four percent of treated adnexa increased in adhesion score, compared to 67% of the untreated controls.

The following products are poly(ethylene glycol) (PEG)-based precursor liquids that are sprayed onto the surgical site and rapidly cross-link on the target tissue to form a flexible, adherent, bioabsorbable gel barrier. They are indicated for use in both open and laparoscopic abdominopelvic surgical procedures. These products are currently available in Europe, but have not been approved in the USA. SprayGel® (Covidien, Waltham, MA) consists of PEG ester and PEG amine precursors, which polymerize within seconds of application on tissue. The reaction scheme is shown in Fig. 5.5. Methylene blue is added to one of the precursor solutions to aid in visualization of the deposition. The applicator is shown in Fig. 5.6 [29]. This gel was approved in Europe in 2001 and costs €275 for a 2×5 mL syringe kit with a laparoscopic applicator. In cases of extensive pelvic surgery, two to five kits may be required to cover the affected area completely [30]. After several days, the adhesion barrier is hydrolyzed into water-soluble PEG molecules and absorbed into the circulatory system, followed by renal excretion.

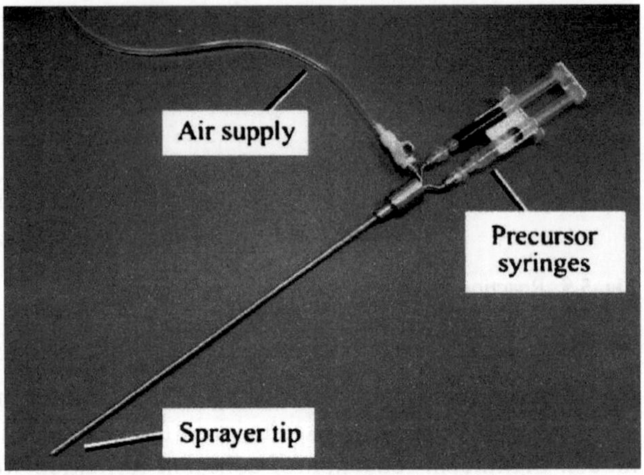

Fig. 5.6 Photograph of a SprayGel applicator [29]

Table 5.2 Adhesion extent and severity scores for a rat cecal abrasion model [29]

	Adhesion extent		Adhesion severity	
Rat	Control	Treated	Control	Treated
1	2	1	1	2
2	2	0	2	0
3	1	0	1	0
4	2	0	2	0
5	0	0	0	0
6	2	0	2	0
7	2	0	2	0
8	2	0	1	0
p value		0.003		0.015

In a preclinical study, rat cecum abrasion and rabbit uterine horn adhesion models were used to evaluate the efficacy of SprayGel [29]. The rat model was created by abrading the cecum with gauze and the opposite abdominal wall with a scalpel, while in the rabbit model uterine horns were abraded with a scalpel. SprayGel was applied onto the sites of injury, and the control group received no treatment. Adhesion formation in rat and rabbit models was evaluated after 12 and 15 days, respectively. The adhesion scoring results are shown in Table 5.2 (rat cecal abrasion model) and Table 5.3 (rabbit uterine horn model). SprayGel significantly reduced the incidence of adhesions in the rat model. Both the extent and severity of adhesions were reduced in the rabbit model.

A porcine adhesion model was developed by laparotomy in which bilateral uterine horn transection and reanastomosis as well as peritoneal sidewall excision were performed [31]. The porcine model allowed the creation of conditions that are relevant to the human surgical environment in terms of organ size and forces exerted on the adhesion barrier. At a second-look laparoscopy within 16 days postoperatively, 90% of the untreated site had adhesions compared with 30% of the SprayGel

Table 5.3 Adhesion extent and severity scores for a rabbit uterine horn model [29]

Rabbit	Adhesion extent		Adhesion severity	
	Control	Treated	Control	Treated
1	3	3	1	1
2	3	1	1	1
3	3	1	1	1
4	3	0	1	0
5	3	3	1	1
6	3	1	1	0.5
7	3	1	1	0.5
8	2	1	1	0.5
9	1	1	1	0.5
10	3	1	1	0.5
p value		0.007		0.023

treated sites ($p=0.006$). The extent and severity of adhesions were significantly lower for the treated site than for the untreated site.

Clinical studies have shown the effectiveness of SprayGel in preventing tissue adhesion. After the completion of myomectomy and before closure, patients were randomly assigned to the treatment or control group [32]. In the treatment group, all suture lines and potentially adhesiogenic surfaces on the uterus and adjacent structures were coated with SprayGel. At 3–16 weeks postoperatively, second-look laparoscopy revealed that 5 of 18 treatment patients were adhesion-free (28%), compared with 1 of 13 (8%) in the control group. Another randomized study was performed on 51 women who underwent open or laparoscopic myomectomy and were treated with or without SprayGel [33]. Seven of 22 patients (31.8%) in the SprayGel group and 2 of 18 (11.1%) patients in the control group were free from adhesion in 3 to 16 weeks after the surgery.

Patients undergoing closure of loop ileostomy were randomized to the group having SprayGel around both limbs of ileostomy for 20 cm or to the untreated control group [34]. Ileostomy was reversed at 10 weeks after construction, and the adhesion formation was evaluated. The SprayGel group had significantly lower overall adhesion scores than the control group (6.11 vs. 9.67; $p<0.0005$). Times taken to mobilize and close the ileostomy were significantly reduced for the treated group.

The next-generation product of SprayGel is called SprayShield™ (Covidien, Waltham, MA), which has been available in Europe since 2008. It consists of PEG ester and trilysine solutions and is chemically identical to Covidien's DuraSeal™ product. With use of low-molecular-weight trilysine, instead of PEG amine, the final gel formed is anticipated to be better mixed and to react better, and thus more adherent than SprayGel. SprayShield completely hydrolyzed within 7 days at 37°C in PBS, while SprayGel took up to 20 days to hydrolyze [35].

Two adhesion barriers, SprayShield and SprayGel, were compared in a laparoscopic porcine model of gynecologic surgery [35]. Eighteen virgin hogs underwent bilateral uterine horn and abdominal sidewall injury via laparotomy. After middle incisions were closed, laparoscopy was performed to apply adhesion barriers, or no

Fig. 5.7 Photograph of Preclude™

application was performed as a control. All animals were evaluated for adhesion formation 10–14 days postoperatively. Both the adhesion barriers demonstrated statistically significant reduction (~50%) in number of adhesions to the site of injury, while only SprayShield demonstrated a statistically significant reduction in adhesion area. The differences between severities of adhesion were not statistically significant.

Adhibit™ adhesion prevention gel (Angiotech Pharmaceuticals Inc., Vancouver, BC) was approved in Europe in 2002 to prevent or reduce postsurgical adhesions during cardiac surgery. It is chemically identical to CoSeal® Surgical sealant. Adhibit™ comes in a 2×4 mL kit. A randomized, controlled, single-blind study was performed on 71 women who had surgery to remove uterine fibroids (myomectomy). Adhesions were measured 8–10 weeks after the surgery, and it was found that 48 women who received Adhibit™ (0.8±2.0) had reduction in adhesions compared with 23 women in a control group who did not receive the gel (2.6±2.2, $p=0.01$) [36].

Sheets

Traditionally, nondegradable membranes such as PTFE film and glutaraldehyde-cross-linked bovine pericardium have been used as adhesion prevention materials. However, their clinical performance is not satisfactory because of the presence of adhesions on suture lines and foreign-body reaction due to their long-term presence. Bioabsorbable materials are therefore preferred for this purpose.

PTFE is a fully fluorinated polymer which is inert. Preclude™ (Gore-Tex, W.L. Gore and Associates) is expanded PTFE (ePTFE), nonabsorbable permanent membrane (Fig. 5.7). For adhesion prevention, fairly large pieces are needed, and they must be sutured into place as it does not adhere to tissue. The capabilities of PTFE

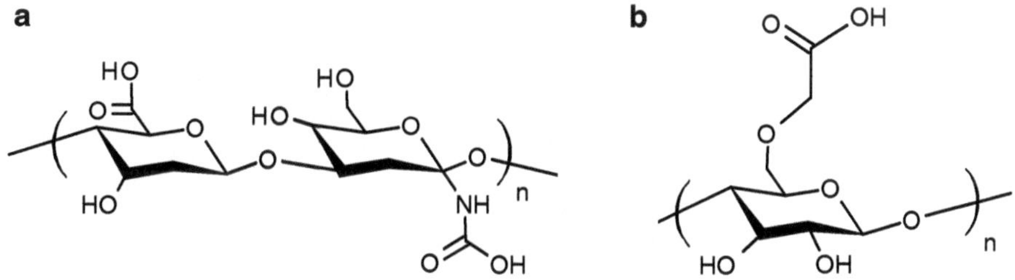

Fig. 5.8 Chemical structure of (**a**) hyaluronic acid (HA) and (**b**) carboxymethyl cellulose (CMC)

membrane (Gore-Tex), PEG hydrogel (SprayGel), and their combination to prevent postoperative pericardial adhesions were compared for patients with hypoplastic left heart syndrome (HLHS) [37]. Eighteen consecutive patients with HLHS were included. The PTFE membrane significantly reduced postoperative adhesions between the heart and the sternum, but with impaired epicardial visibility. SprayGel did not prevent formation of pericardial adhesions.

Clinical studies have exhibited the effectiveness of this PTFE product. A study involved 27 women undergoing myomectomy with at least two incisions on the uterine fundus and posterior uterine wall of similar length [38]. At laparotomy, the two incision sites were randomly assigned to be covered with PTFE or left untreated. Second-look laparoscopy after 2–6 weeks after the myomectomy revealed that 55.6% of PTFE-covered sites and 7.4% of untreated sites were completely free of adhesions. The severity of adhesions at the PTFE sites was significantly lower than at the control sites. Another study compared PTFE and Interceed® in the prevention of postsurgical adhesions for 32 women with bilateral pelvic sidewall adhesions undergoing reconstructive surgery [39]. The sidewalls covered with PTFE had a significantly lower incidence of adhesions (85%) compared to those covered with Interceed® (65%). Despite the effectiveness of PTFE, its usage is limited due to the need for a removal procedure a few days after application. It is also difficult to apply by laparoscopy. However, one study suggested that the removal of PTFE membrane is not required. Long-term safety of PTFE was investigated in a clinical study involving 146 patients in whom the membrane was implanted permanently during peritoneal reconstruction [40]. The mean postoperative observation time was 3.5 years. A single case of postoperative infection was observed, but did not require removal of membrane, and results for all other patients were good.

Most commonly used bioabsorbable films for adhesion prevention are Seprafilm® (Genzyme Corporation, Cambridge, MA) and Interceed® (Johnson & Johnson, New Brunswick, NJ). Seprafilm® is a transparent membrane, $13 \times 15\ cm^2$ in size, which is composed of sodium hyaluronate and CMC (Figs. 5.8 and 5.9). This film adheres to sites of injury by absorbing water from the surrounding region and turns into a hydrophilic gel that separates opposing tissues during the postoperative healing phase. Seprafilm® is degraded and disappears in 7 days. Clinical studies have demonstrated the effectiveness of Seprafilm® [41], but there are concerns regarding a

Fig. 5.9 Photograph of Seprafilm®

higher incidence of anastomotic leaks in cases in which the film is placed directly around the anastomosis. The use of Seprafilm® in this location is prohibited by instructions provided by the manufacturer.

The efficacy of Seprafilm® as an antiadhesive barrier for bowel anastomoses was investigated using a rabbit model [42]. Sixty-four rabbits underwent complete or partial (90% anastomosis to simulate anastomotic leakage) large bowel anastomosis, and half of them were treated by wrapping Seprafilm® while the other half were untreated controls. The average anastomotic bursting pressure did not differ significantly between the two groups at 7 and 14 days.

In preclinical studies, Tanaka et al. investigated the efficacy of Seprafilm® in preventing pleural adhesion in a rat model [3]. An adhesion model on rat pleura was created using a combination of mechanical, chemical, and ischemic injury. The parietal pleura beneath the ribs on both sides of the thoractomy incision was thoroughly scraped with a bone curette, and then the surface of the left lung was abraded with five strokes with gauze soaked with 0.1 mL of tincture of iodine. Finally, the chest wall was kept ischemic by electrocautery applied to the intercostal arteries on both sides of the thoracic incision. Seprafilm® was inserted between the lung and the parietal pleura, and the control underwent infusion of saline alone. On the ninth postoperative day, the severity of adhesion formation was examined. The length of adhesion in the Seprafilm® group was 3.5 ± 3.5 mm, which was significantly smaller than that in the control group (14.7 ± 4.7 mm, $p<0.05$). The severity of adhesion was also lower in the Seprafilm® group. They measured the activity of type 1 plasminogen activator inhibitor (PAI-1) in the intrapleural lavage fluid (ILF) at 24 h postoperatively since this has the strongest influence on adhesion formation after surgery. PAI-1 activity was 23.37 ± 2.57 U/mL in the saline-treated group and 17.85 ± 3.06 U/mL in the Seprafilm® group, indicating that it inhibits postoperative adhesion formation through significant repression of PAI-1 activity.

The efficacy of Seprafilm® was evaluated using a rat cecal abrasion or sidewall injury model [43]. For the cecal abrasion model, rat cecum was isolated and abraded with special gauze with constant abrasion force and without hemostasis,

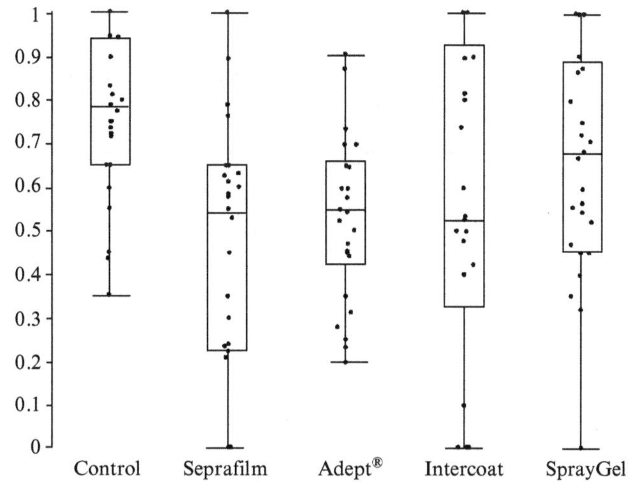

Fig. 5.10 Boxplot showing the fractions of injured areas that had adhesions. The minimum value, lower quartile, median, and upper quartile are indicated [44]

Seprafilm® was placed on the site of injury, and the incidence and severity of adhesions were evaluated 7 days later. Animals treated with membrane exhibited significantly less incidence of adhesions even in the presence of blood compared to the nontreated group.

Due to differences in preclinical models and methods of adhesion assessment, direct comparison of products available today is lacking, and the results in the literature cannot easily be compared. Notwithstanding this, Rajab et al. directly compared four commonly used adhesion barriers against a control group in a rat model [44]. Standardized lesions were created in Wistar rats using electrocautery and suturing. The adhesion model was treated with Seprafilm®, Adept®, Intercoat®, SprayGel, or no barrier ($n=30$). Results were observed 2 weeks postoperatively. Figure 5.10 shows the fraction of the traumatized areas covered by adhesions. The adhesion-free incidence was 20% ($n=6$) for the Seprafilm® group, 20% ($n=6$) for the Intercoat® group, 3% ($n=1$) for the SprayGel group, and 0% for the Adept® or the control group. It was concluded that despite the observed statistically significant differences between different barriers, a significant adhesion problem remained for all of the tested barriers.

Clinical studies have shown that Seprafilm® reduces adhesion formation. One hundred and eighty-three patients with ulcerative colitis or familial polyposis who were scheduled for colectomy and ileal pouch-anal anastomosis with diverting-loop ileostomy were randomly assigned to receive or not receive Seprafilm® placed under the midline incision before abdominal closure [45]. Eight to 12 weeks later, laparoscopy was used to evaluate formation of adhesions to the midline incision for 175 patients who were assessable. Seprafilm® significantly reduced the incidence and severity of adhesions, as shown in Table 5.4. In the control group (90 patients), only five patients had no adhesions (6%), whereas 43 of 85 patients in the Seprafilm

Table 5.4 The incidence, extent, and severity of postoperative adhesion formation to the midline incision [45]

	Control group ($n=90$)		HA membrane group ($n=85$)		
	n	Percent	*n*	Percent	*p* value
Incidence					
No adhesions	5	6	43	51	< 0.00000000001[a]
Adhesions	85	94	42	49	
Extent[b]					
All patients	90	63 ± 34	85	23 ± 34	< 0.0001[c]
Patients with adhesion	85	67 ± 31	42	48 ± 34	= 0.0008[c]
Severity[d]					
No adhesions	5	6	43	51	< 0.001[e]
Grade 1	4	4	12	14	
Grade 2	29	32	17	20	
Grade 3	52	58	13	15	

[a]Fisher's exact test
[b]The proportion of the total length of the initial surgery midline incision associated with any adhesions, as determined by dividing the length associated with adhesions (cm) by the overall length of the initial midline incision (cm). Data are reported as mean ± standard deviation
[c]Student's *t* test
[d]Grade 1, filmy thickness, avascular; Grade 2, moderate thickness, limited vascularity; Grade 3, dense thickness, vascularized
[e]Wilcoxon's rank sum test
n number, *HA* sodium hyaluronate

group were free of adhesions (51%). The adhesions involved most of the posterior aspect of the midline incision and arose from the omentum, small bowel, and exposed peritoneal surfaces. Adhesions to the omentum were reduced from 79% in the control group to 39% in the Seprafilm group.

Kusunoki et al. evaluated the effectiveness of Seprafilm in preventing abdominal adhesions after radical resection of rectal carcinoma and examined whether Seprafilm had any adverse effects in patients treated with radiotherapy and chemotherapy [46]. Sixty-two patients participated and received preoperative radiotherapy followed by a two-stage operation and 5-fluorouracil (5-FU)-based systemic chemotherapy. At the time of ileostomy closure, the severity and extent of adhesions were evaluated. No adhesion under the midline incision was observed for 26 patients in the Seprafilm group ($n=30$, 86.7%) and 4 patients in the control group ($n=29$, 13.8%). In the case of peristromal adhesions, 4 and 0 patients had no adhesions in the Seprafilm (13%) and control (0%) groups, respectively. The total scores for severity and extent of adhesions are shown in Table 5.5. Seprafilm significantly reduced adhesions in both the midline incision area and peristromal area. This in turn reduced operative time, blood loss, and extent of the incision at ileostomy closure. There was no association of Seprafilm with any postoperative complications, chemoradiation-related toxicity, recurrence, or survival rates.

The efficacy of Seprafilm in preventing postoperative small-bowel adhesive obstruction was evaluated after distal gastrectomy (DG) [41, 47]. Of 282 patients

Table 5.5 Degree of adhesion formation expressed as the percent of investigational surgical site (ISS) showing adhesion as assessed during the second sternotomy [46]

Percent of ISS	Barrier	Control	*p* value
Grade 0: no adhesions	1.1±3.84	1.0±4.67	0.9143
Grade 1: mild adhesions	32.2±35.94	15.0±26.38	0.0065
Grade 2: moderate adhesions	45.6±36.40	34.5±35.64	0.1228
Grade 3: severe adhesions	21.1±36.84	49.5±42.66	0.0005

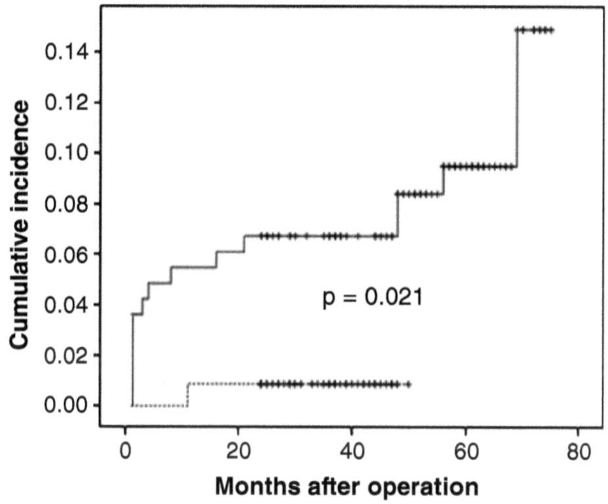

Fig. 5.11 Graph showing the cumulative incidence of small-bowel adhesive obstruction after distal gastrectomy. *Solid line* = no Seprafilm and *dashed line* = Seprafilm [41]

diagnosed with gastric cancer who underwent open DG with Billroth I anastomosis, Seprafilm was used for 113 patients [41]. Figure 5.11 shows the cumulative incidence of adhesive obstruction after surgery. Significantly lower incidence was observed in the Seprafilm group than in the non-Seprafilm group ($p = 0.021$). Two years postoperatively, the incidences of adhesive obstruction were 0.9% and 6.5% for the Seprafilm and control groups, respectively.

Hayashi et al. performed a randomized clinical trial of Seprafilm to reduce postoperative small-bowel obstruction in 144 patients who underwent gastrectomy [47]. A slightly but not significantly lower cumulative incidence of adhesions was observed in the Seprafilm group ($n = 70$) compared to the control group ($n = 74$). This trial included not only distal gastrectomy but also total gastrectomy, which might have a higher risk of adhesive obstruction.

Large-scale clinical studies were conducted to evaluate the efficacy of Seprafilm [48, 49]. The efficacy of Seprafilm in reducing adhesion-related postoperative bowel obstruction after abdominopelvic surgery was evaluated in a total of 1,791 patients, most of whom had inflammatory bowel disease [48]. The patients were randomly divided into a Seprafilm group ($n = 882$) and a nontreated group ($n = 909$), and complications observed within the first month after surgery were evaluated. The safety of Seprafilm was confirmed with respect to abdominal abscess, pelvic abscess, and pulmonary embolism. The subpopulation of patients for whom Seprafilm was wrapped around a fresh bowel anastomosis experienced more frequently leak-related events that included anastomotic leakage, fistula, peritonitis, abscess, and sepsis ($p \leq 0.05$). Accordingly, it is recommended that wrapping the suture or staple line of a fresh bowel anastomosis with Seprafilm be avoided.

The safety and long-term clinical benefit of Seprafilm in reduction of adhesive small-bowel obstruction after intestinal resection were evaluated with a follow-up period of 3.5 years [49]. One thousand seven hundred and one patients who underwent intestinal resection were randomized to receive Seprafilm before closure of the abdomen ($n=840$) or without treatment ($n=861$). The incidence and type of bowel obstruction were compared between the two groups. The overall rate of bowel obstruction exhibited no difference between the treatment and control group, but the incidence of adhesive small-bowel obstruction requiring reoperation was significantly lower for Seprafilm patients than for no-treatment patients (1.8% vs. 3.4% ($p<0.05$)).

Seprafilm was also clinically evaluated for reduction of pericardial adhesions. In congenital cardiac surgery, reoperation is not uncommon. Surgical adhesions lead to more difficult sternal reentry and cardiac dissection, to blunted visibility of distinct cardiac structures, to potential injury of cardiac structures, as well as to an increased risk of surgical bleeding, all associated with an increase in morbidity and mortality. Walther et al. applied Seprafilm to 350 of 1,024 patients after surgery for congenital heart defects [50]. Among these, 30 patients underwent reoperation and were evaluated in comparison to ten randomly reoperated patients. Remarkable reductions in the tenacity score (3.3 vs. 4.3) and in the extent of adhesions (77.7% vs. 86%) were observed for the Seprafilm group compared to the control group.

Naito et al. evaluated the efficacy of Seprafilm in combination with ePTFE or autologous pericardium to reduce pericardial adhesions in an experimental pericardial adhesion model in 24 beagle dogs [51]. During operation, pericardium was opened, and the exposed epicardial surfaces of the right ventricular outflow tract were desiccated and abraded for 3 min with gauze. Then, the pericardium was resected to the size of samples and closed. Four groups were tested (each group, $n=6$): Group A: ePTFE alone, B: Seprafilm + ePTFE, C: autologous pericardium only, and D: Seprafilm + autologous pericardium. Pericardial adhesions were evaluated at necropsy 4, 8, and 12 weeks postoperatively. Figure 5.12 shows the adhesion tenacity scores and tissue thickness. Combining the Seprafilm with ePTFE or autologous pericardium yielded significant reduction in pericardial adhesion formation.

Despite the success in reducing adhesion using Seprafilm in clinical and preclinical studies, handling of this film is quite difficult because of its poor mechanical properties and brittleness when wet. Placing this film through a laparoscope is very difficult since it easily breaks when bolted. Replacing the film on wet tissue is also difficult.

Another commonly used bioabsorbable sheet is Interceed® (Ethicon Inc., Somerville, NJ) which is a knitted fabric of oxidized regenerated cellulose with a size of $7.6 \times 10.2\ cm^2$ (Figs. 5.13 and 5.14). It turns into a gel when placed on tissues and is absorbed in 2 weeks. Interceed® has been available since 1990 for applications

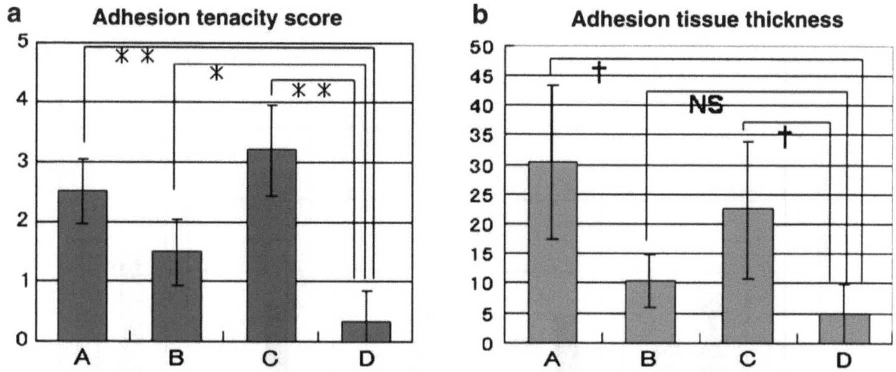

Fig. 5.12 Histogram showing the (**a**) mean adhesion tenacity score and (**b**) mean adhesion tissue thickness. *A* ePTFE alone, *B* Seprafilm+ePTFE, *C* autologous pericardium only, *D* Seprafilm+autologous pericardium. *$p<.05$ and **$p<.01$, respectively (Scheffé's test). †$p<.01$ (Fisher's PLSD test), *NS* not statistically significant [51]

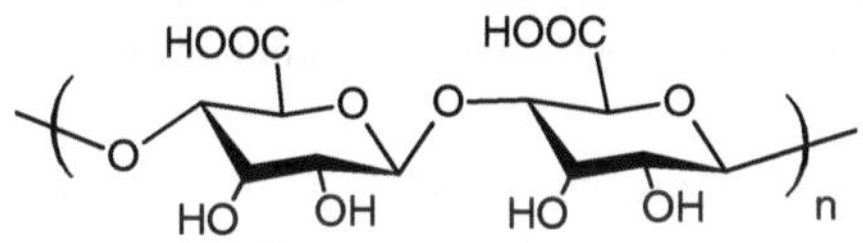

Fig. 5.13 Chemical structure of oxidized cellulose

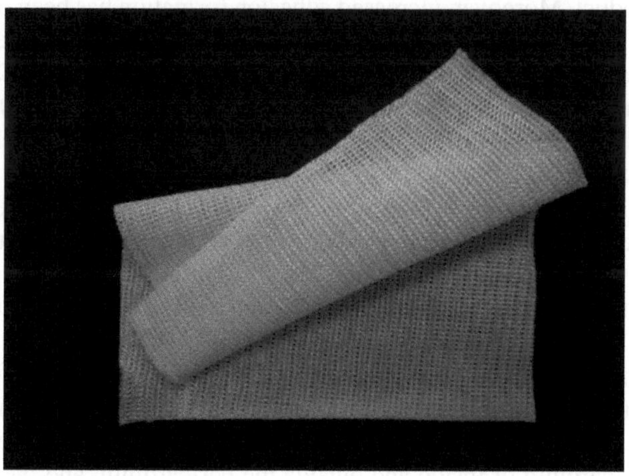

Fig. 5.14 Photograph of Interceed®

in gynecologic pelvic surgery. Unlike Seprafilm, Interceed® is easy to handle and can be placed via laparoscopy. It has been reported that Interceed® prevents tissue adhesion effectively when blood contamination is avoided during application of it. A meta-analysis of ten randomized, controlled studies revealed 24.2% reduction of

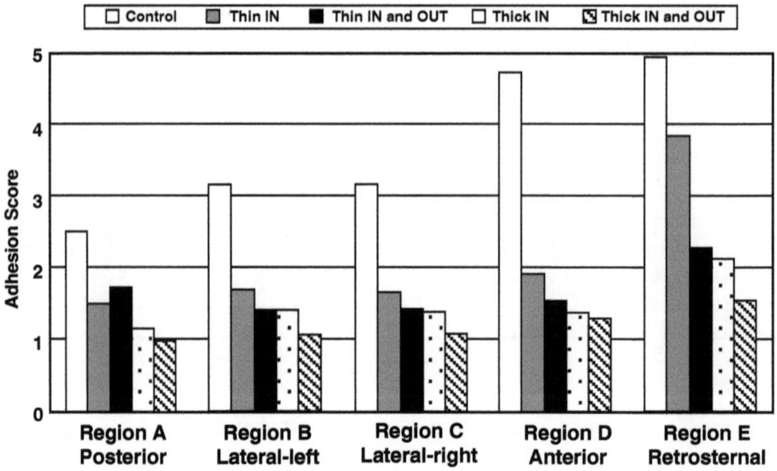

Fig. 5.15 Summary of the adhesion scores (ranging from 0 to 5) for the different anatomical regions of the heart 4 weeks postoperatively. *Thin* 20 μm thickness film, *Thick* 50μm thickness film, *IN* inside the pericardium, *OUT* outside the pericardium [54]

adhesion formation on the Interceed®-treated site, compared to the untreated control site [52]. Use of Interceed® resulted in a greater reduction in extent of adhesion (1.1 ± 0.4 cm^2) than good surgical technique alone ($p < 0.001$). However, as complete blood evacuation is not always possible in clinical situations, surgeons are not keen to use this product. Moreover, increased adhesion formation has been observed when Interceed® is applied to areas where blood accumulation cannot be prevented [53].

CardioWrap® and SurgiWrap® (MAST Biosurgery USA Inc., San Diego, CA) are translucent membranes composed of L-lactide (70) and D,L-lactide (30) and are used for adhesion prevention barriers in cardiovascular and abdominal/pelvic surgeries, respectively. Sheets of 5×7 cm^2 (SurgiWrap® only), 10×13 cm^2, or 13×20 cm^2 are available. Thickness comes in two sizes, 20 and 50 μm, which affect the strength retention time. The membrane can easily be repositioned even in a wet environment. More than 80% of the copolymer remains after the initial 8 weeks but is fully absorbed within 6 months. The molecules are slowly hydrolyzed into carbon dioxide and water and finally released from the body through the lungs. This film can be sutured.

In a preclinical study, an adult pig model (total 25 pigs) was used for retrosternal adhesion formation via inferior hemisternotomy to evaluate the efficacy of these polylactide films [54]. Two thicknesses of films (20 and 50 μm) were placed either inside the pericardium or inside and outside the pericardium, and the control group was untreated. After 4 weeks, all animals demonstrated adhesions between the posterior and lateral surfaces of the heart and pericardium. However, as shown in Fig. 5.15, polylactide films reduced the severity of adhesions in all anatomical regions examined.

A prospective control clinical study was performed to assess the efficacy of SurgiWrap® in preventing adhesion formation in patients undergoing a major colorectal surgery [55]. After the surgery, SurgiWrap® was placed under the midline

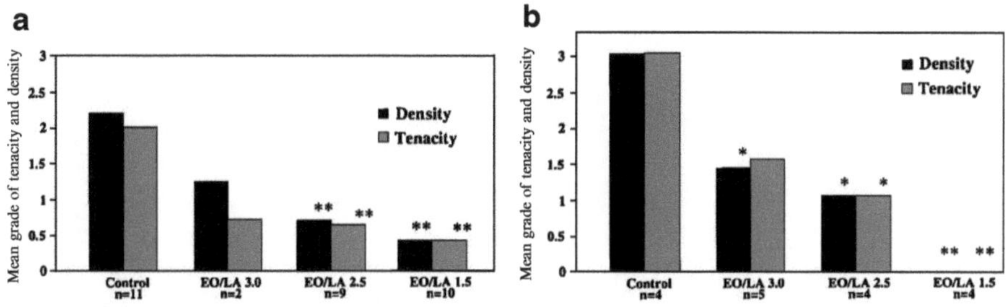

Fig. 5.16 Histograms showing the mean adhesion scores for tenacity and density for each group in models 1 and 2. *EO* poly(ethylene glycol), *LA* poly(lactic acid) [56]

incision just before closing the abdominal wall for ten patients. The control group (Nine patients) received no treatment. After 14 weeks postoperatively, the patients were admitted for laparoscopic ileostomy closure, and during that time, adhesion formation was evaluated. In the SurgiWrap® group, five patients had no adhesion beneath the midline incision, in contrast to 0 patients in the control group. The severity of adhesions was significantly lower for the treated group than for the control.

REPEL-CV® (SyntheMed Inc., Iselin, NJ) is composed of a blend of poly (lactic acid) (52%) and poly(ethylene glycol) (47%). It has been commercially available in Europe since 2006 and was approved by the FDA in 2007. Its size is $18 \times 13.5\ cm^2$, and its thickness is 137 μm. In preclinical animal studies using rabbit and canine models, three films with various poly(ethylene glycol) and poly(lactic acid) ratios (EO/LA ratios of 1.5 (REPEL-CV®), 2.5, and 3.0) were tested. All the films were found to reduce adhesion formation significantly. In a canine model, after desiccation and abrasion of the epicardium, film was placed over the heart [56]. Pericardium was closed to allow intrapericardial adhesions or left open and attached to the chest wall to induce retrosternal adhesion. Adhesion formation was evaluated after 3 weeks. As shown in Fig. 5.16, the mean tenacity and density scores of adhesions were significantly lower for treated groups than for the control. The film with an EO/LA ratio of 1.5 was found to be optimal.

A randomized study was performed to evaluate the efficacy of REPEL-CV® in 142 infants undergoing initial sternotomy for eventual staged palliative cardiac operations. Before chest closure, patients were randomized to barrier film placement ($n = 54$) or control ($n = 49$) [57]. Figure 5.17 shows an illustration of the technique of barrier film implantation. At the time of chest closure, the barrier was applied over the epicardial surface of the heart and fixed in place with four interrupted poly(glycolic acid) sutures. In the control group, the chest was closed without the use of any adjuncts. The extent and severity of adhesions at the investigational surgical site (ISS) were evaluated at repeated sternotomy 2–13 months later. As shown in Fig. 5.18, significantly fewer patients in the REPEL-CV® group had any severe adhesions compared with the control group. There were no statistically significant differences in adverse events between the two groups, indicating no adverse events due to REPEL-CV®.

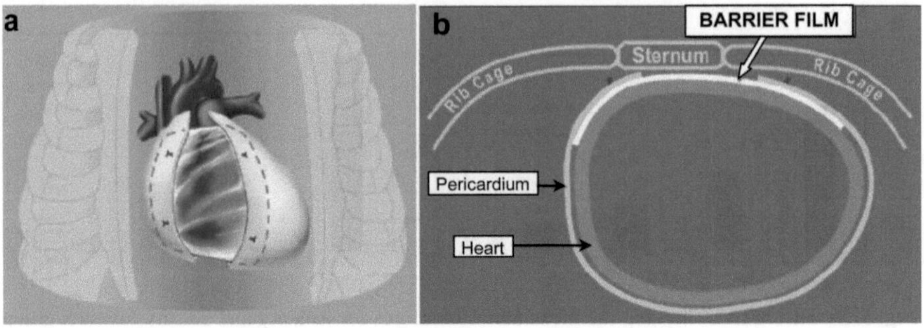

Fig. 5.17 Schematic illustration of the technique for barrier film implantation. (**a**) A frontal view showing the film overlapping the pericardial edge. (**b**) An axial diagram showing the positions of the sternum, pericardium, film, and heart after implantation [57]

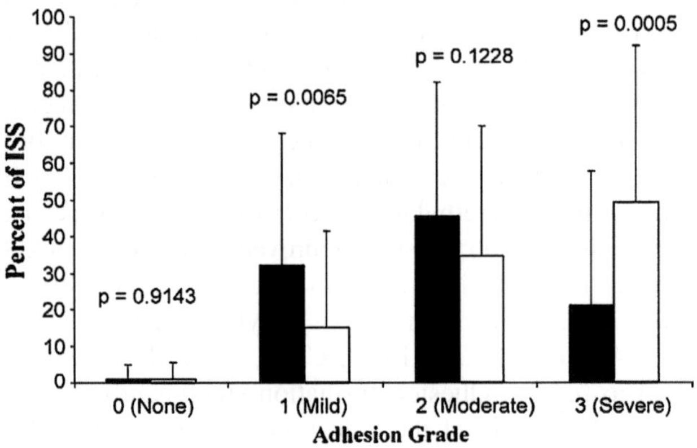

Fig. 5.18 Histogram showing the distribution of adhesions based on severity grade between the barrier (*black bars*) and control (*white bars*) groups [57]

Cova CARD™ and Cova ABDO™ (Biom'Up, Lyon, France) are CE marked patented membranes that are made of purified porcine type I collagen membrane. The membrane is cross-linked with an oxidized polysaccharide [58]. Dry membrane has a thickness of about 100 μm. When degraded by proteolysis, fragments are eliminated through the kidneys, and the film disappears within 6 months. In a preclinical study, the efficacy of Cova CARD™ was compared with that of Seprafilm [59]. Sixteen sheep underwent a sternotomy followed by scratching of the heart surface. They were then divided into three groups: pericardium left opened ($n=4$), placement of Seprafilm ($n=6$), or Cova CARD™ ($n=6$). When the intensity of adhesion was assessed 4 months postoperatively, Cova CARD ™ was found to have been almost totally absorbed, and the adhesion score was significantly lower than that of the Seprafilm group (1 vs. 3 (score from 1, absence to 3, tight)). There was no inflammatory reaction, and the extent and density of fibrosis were small.

Three types of antiadhesive films (SurgiWrap®, Prevadh®, and Seprafilm®) were compared for intraperitoneal onlay mesh hernia repair in a rat model [60]. Prevadh®

(Covidien, Mansfield, Massachusetts) is made of oxidized bovine atelocollagen type 1, poly(ethylene glycol), and glycerol, which is degraded within 3 weeks. One macroporous polypropylene mesh (2×2 cm^2) was fixed with four nonabsorbable sutures on the abdominal wall of rat, and a thin layer of fibrin sealant (0.2 mL) was applied with the spray system. A 2.5×2.5 cm^2 patch of antiadhesive film was placed on the mesh, and the abdominal wall was closed. After 30 days, rats were killed, and adhesion formation was assessed. The control without barrier exhibited severe adhesions in all animals. The use of Prevadh and Seprafilm significantly reduced adhesion formation at 30 days after surgery. On the other hand, SurgiWrap® did not have a significant antiadhesive effect, and its degradation was faster than predicted from the manufacturer's information (6–8 weeks). SurgiWrap® remained stiff when hydrated and appears inadequate for laparoscopy.

Recent Studies

Fibrin-Based Material

Fibrin sealant has the advantage of decreasing bleeding and increasing the production of plasminogen activator and plasminogen activator inhibitor-1, which may be beneficial in the prevention of adhesion [61]. In a rat model, intraperitoneal adhesion formation was induced by excision of 1×3 cm^2 of the peritoneal parietal muscular layer, followed by closure with interrupted 3–0 silk sutures [62]. At 1 week postoperatively, the median length of adhesions in the defects covered with fibrin sealant applied by a syringe (10.5 mm) or by spray (14 mm) was less than that of the control group (25 mm) ($p < 0.001$). Fibrin glue has also been shown to reduce adhesion formation onto permanent prosthetic materials such as polypropylene and PTFE in rat and pig models [63–65].

Prevention of adhesion by fibrin glue on a rabbit flexor tendon adhesion model was examined by Feykman et al. [66]. A partial laceration was produced and closed with 5–0 Dexon sutures, and covered with fibrin glue for the experimental group. The operated paw was then immobilized by flexion of the wrist for 3 weeks, or left free for voluntary weight-bearing immediately after surgery. Although the immobilized tendons exhibited no difference in adhesion formation at 6 weeks, the mobile tendons exhibited less adhesion formation in the fibrin glue group than the control. Clinical data on this application are scarce, and fibrin glue is not approved for this surgical purpose [67].

Gelatin-Based Materials

Gelatin can be cross-linked by thermal, UV, or chemical means, and its degradation can be controlled easily by modifying the degree of cross-linking. UV cross-linked gelatin films were investigated for use as an antiadhesive material [13]. As shown in

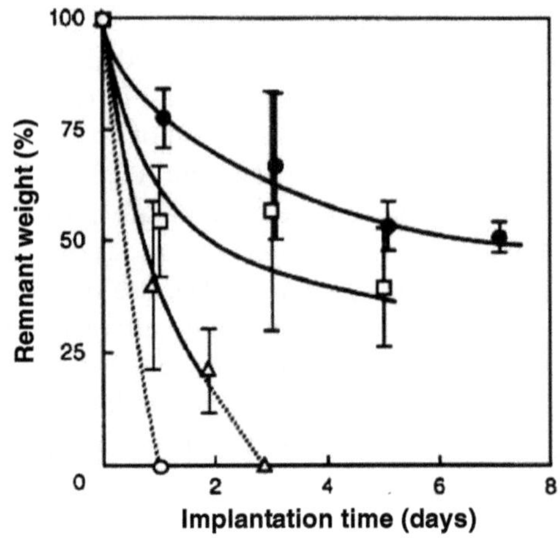

Fig. 5.19 Graph showing in vivo degradation of cross-linked gelatin films with increasing UV irradiation time. Films irradiated for 5 and 10 h could not be seen with the naked eye after 1 and 3 days, respectively. Irradiation time: *open circle* = 5 h, *triangle* = 10 h, *square* = 20 h, and *filled circle* = 40 h. *Error bars* indicate the standard deviations [13]

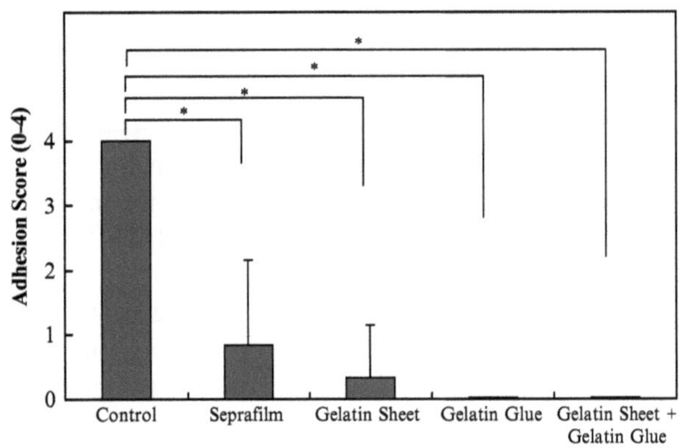

Fig. 5.20 Adhesion scores for peritoneal adhesions in the rat model. $*p<0.01$, $n=6$ [68]

Fig. 5.19, the in vivo degradation of cross-linked gelatin film could be controlled by UV irradiation time. When gelatin films with different degradation rates were evaluated for antiadhesion efficacy using a rat adhesion model, it was found that the film which was degraded in 3 days exhibited the best antiadhesive properties, whereas longer-lasting films were less efficient, probably due to the prolonged presence of the films inducing foreign-body reaction and resulting in tissue adhesion.

A thermally cross-linked gelatin sheet consisting of sponge and film layers was developed. The sponge layer was provided to absorb fluid and adhere to the body tissue, and the film layer to prevent invasion of undesired cells and tissues and avoid adhesion formation. Even in the presence of oozing-type bleeding from the surgical site, the sheet could be applied for astriction, and it did not adhere to the gauze or gloves due to the film layer. The efficacy of the sheet for adhesion prevention was investigated using a rat adhesion model, and the severity of adhesion was scored. As shown in Fig. 5.20, the sheet significantly reduced adhesion score compared to the untreated control. Antiadhesive properties of gelatin glue consisting of 23 wt.% gelatin and 0.12 wt.% GA was also evaluated using a rat adhesion model [68].

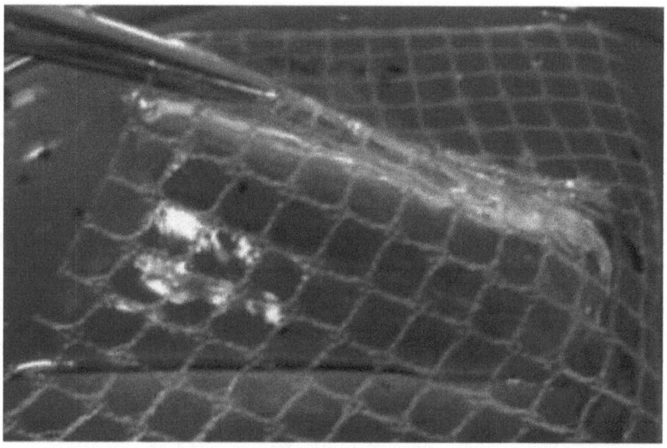

Fig. 5.21 Photographic image of the wet gelatin sheet reinforced with PGA

Gelatin glue was applied to the rat cecum abrasion site with or without the previously described gelatin sheet, and adhesion was evaluated 2 weeks postoperatively. All rats treated with gelatin glue with or without gelatin sheet showed no adhesion formation (Fig. 5.20).

Formation of adhesions between the heart and the chest wall after open heart surgery can cause significant problems such as bleeding, injury to the heart, or great vessels, and longer operative and perfusion times. The most commonly used pericardial substitute, PTFE, has problems such as severe inflammation and diffuse fibrosis. To address these problems, gelatin sheet was studied as a pericardial substitute [69, 70]. A thermally treated gelatin sheet, which exhibited degradation in 4 weeks, exhibited reduced formation of pleural and pericardial adhesion and inflammatory reaction, compared with the PTFE sheet in a canine model [70]. Despite the effectiveness of the gelatin sheet as a pericardial substitute, there are two considerable drawbacks in its structural properties in the hydrated state: one drawback of the sheet is that it is too malleable to handle, and the other is that it is too low in tensile strength so that it cannot steadily hold sutures to the native pericardium.

To overcome these problems, the gelatin sheet was reinforced with bioabsorbable PGA mesh, as shown in Fig. 5.21 [69]. The modified sheet exhibited tenfold higher tension of disruption at the suturing margin than the unmodified gelatin sheet (619 ± 141 vs. 62 ± 7 gf). This is similar to that of PTFE. The effectiveness of this reinforced sheet in the prevention of adhesion was confirmed in a canine model.

Collagen-Based Materials

A collagenous device based on amniotic membrane was documented for adhesion prevention. Young investigated the use of trypsin-treated, gamma-irradiated human amniotic membrane in preventing postoperative adhesions in a rabbit uterine horn adhesion model [71]. At 30 days postoperatively, only 4 (13.4%) of 30 uterine horns treated with amniotic membrane exhibited any adhesion formation, whereas all

24 untreated controls exhibited severe adhesion formation at the site of injury. Miyata and colleagues [72] studied the efficacy of amniotic membrane with slow release of heparin in a dog peritoneal adhesion model. The human amnion was cross-linked with a polyepoxy compound after the impregnation of protamine sulfate, and was heparinized. The membrane showed an excellent antiadhesive effect in all dogs, whereas the untreated control exhibited severe adhesions. On the other hand, Badawy et al. [73] found that the amniotic membrane was ineffective in preventing postoperative adhesions in a rat model. These conflicting results may have arisen due to use of tissue-based material, which varies between samples. On the other hand, the use of soluble collagen allows uniform samples to be prepared.

The efficacy of collagen membrane made from pepsin-soluble calf skin collagen in preventing peritoneal adhesion formation in a rat model was evaluated [74]. To synthesize the membrane, fibrillar collagen was reconstituted from the solution by precipitation using a water-soluble polymer, PEG 4000, to form tactoids. These were then allowed to further aggregate to form larger, fibrillar structures. After drying, the membrane was 0.18–0.20 mm thick, with a density of 36–40 g/m^2. Adhesion formation in a rat model was evaluated at 28 days postoperatively. The collagen membrane performed significantly better than either Interceed® or the untreated control group.

Collagen gel and film have been produced commercially for antiadhesion barriers. Collagen gel (Imedex Biomateriaux, Chaponost, France) is based on a highly purified bovine atelocollagen treated with periodic acid. The gel is supplied in a dual-syringe system containing 2 mL of acidic collagen solution (15% collagen and 5% PEG at pH of 3.2) and 0.58 mL of buffer (0.2 M phosphate/0.3 M carbonate at pH of 10.5). The dual-syringe system is heated in a special heater provided by Imedex Biomateriaux to 60°C to obtain the appropriate viscosity for mixing. Mixing the two solutions neutralizes the collagen solution and causes it to form an adherent gel. Collagen film (Imedex Biomateriaux) is prepared from a diluted, neutralized solution of the gel containing the same proportions of collagen and PEG (3:1) as the collagen gel. The diluted solution is air-dried under sterile airflow to obtain a thin, transparent film (0.133 g/cm^2).

Arnold et al. [75] evaluated the collagen gel and film for prevention of surgical adhesions in a rat model and compared with Seprafilm and fibrin glue (Tisseel™). Surgical defects were created on the right abdominal wall and cecum of rat, and samples were applied either to both wounds on the wall and cecum or only to the wall wound. The control group received no treatment. Seven days postoperatively, adhesions were examined, and the results are shown in Table 5.6. Both the collagen gel and film effectively reduced the incidence of adhesion formation. Use of fibrin sealant resulted in adhesion formation in 9 of 10 animals, similar to that of the control. However, the strength of adhesions formed was significantly weaker than that in the control group.

Gel Amidon Oxydé (GAO, Imedex Biomateriaux) is another collagen sealant based on the chemical cross-linking of collagen and oxidized starch. When mixed, gel is formed in 1–2 min. The efficacy of the gel in preventing epidural adhesions in a rat laminectomy model was investigated [76, 77]. It was found that the gel effectively reduced adhesions without causing side effects in a rat model during the first 3 months postoperatively.

Table 5.6 Incidence and magnitude of adhesion formation at the 7th day following surgery [75]

Treatment	Incidence of adhesion (n/n)	Area of adhesion (cm^2)	Surface with defect involved (%)	Mean maximal strength of the formed adhesions (N)
Collagen gel (both defects)	3/23*	0.17±0.10*	9±5*	0.45±0.27*
Fibrin sealant (both defects)	9/10	1.02±0.37	51±18	0.49±0.18*
Collagen gel (AWD only)	0/10*	–	–	–
Collagen film (AWD only)	3/10*	0.12±0.06*	6±3*	0.67±53*
Seprafilm (AWD only)	2/10*	0.85±0.30	43±15	0.93±0.28
None (controls)	34/35	1.27±0.47	64±23	1.30±0.43

Note: Values are expressed as mean ± standard deviation unless otherwise indicated. *AWD* abdominal wall defect, *SH/CMC* sodium hyaluronate/carboxymethyl cellulose
*$p < 0.01$ (vs. controls)

Hyaluronic Acid–Based Materials

Hyaluronic acid (HA), which is a glucosaminoglycan, is an important component of ECM of the peritoneal epithelium and is produced by peritoneal epithelial cells. HA has been widely used in medical applications. It has been reported that exogenous HA significantly inhibits fetal fibroblast proliferation, suggesting a role of HA in preventing fibrosis and scar formation in fetuses as well as in the early phase of wound healing [78–80]. HA in solution form dissipates rapidly with time and is not effective in preventing adhesion formation. Cross-linking of HA can be performed physically, ionically, or covalently, and various forms of HA including viscoelastic gels and solid films have been investigated and shown to be effective to prevent adhesions.

Auto-cross-linked polysaccharides (ACP, Fidia Advanced Biopolymers Srl, Abano Terme, Italy) are HA derivatives obtained from the cross-linking process which does not introduce any foreign bridge molecules [81]. The carboxyl groups of HA were esterified with hydroxyl groups belonging to the same molecule and/or different molecules of HA, resulting in the formation of a mixture of lactones and intermolecular ester bonds. The level of cross-linking can be varied by modulating the reaction conditions. No bridge molecules are present between the auto-cross-linked HA chains, and the natural components of HA are thus released upon degradation of ACP. The cross-linked HA has an increased viscosity and extends residence time in vivo, while still possessing the tolerability and biocompatibility properties of the original polymer.

The viscoelastic properties of ACP were investigated by varying the cross-linking density and concentrations [81]. At fixed concentration, elasticity was substantially increased by increasing the level of chemical cross-linking. The in vivo pharmacokinetic behavior of ACP gel was investigated after intraperitoneal administration of tritium-labeled ACP gel to beagle dogs [82]. It was found that the ACP administered into the peritoneal cavity was eliminated very slowly by active initial catabolism at the injection site and that the products of catabolism were only CO_2 and H_2O, confirming the same metabolic pathway for native unmodified hyaluronan. The anti-adhesive effects of ACP gel of different concentrations (1, 2, 4, and 6%) were investigated using a rabbit abdominal adhesion model involving the creation of defects in the parietal peritoneum and muscular fascia and cecal abrasion [83]. Ten weeks after surgery, the percentage of adhesion-free animals in the control group was 15%, but 60, 84, 90, and 84% in the 1, 2, 4, and 6% ACP gel concentration groups, respectively ($p=0.001$). The mean adhesion scores in the ACP gel group were also significantly reduced compared to that in the control group.

The efficacy of ACP gel in postsurgical adhesion prevention during bleeding or inadequate hemostasis was evaluated in a rabbit model [84]. Exudative abdominal bleeding (oozing model) and severe bleeding (bleeding model) were created on the uterine horn of rabbits. Three different concentrations of ACP gels (2, 4, and 6%) or Seprafilm were applied without attempting hemostasis. Three weeks postoperatively, all the ACP gel and Seprafilm groups significantly reduced adhesion formation in both the models, compared to that in the untreated control. No statistical differences were observed between the different treatment groups.

Vorevolakos et al. [85] synthesized ferric ion-cross-linked networks of HA based on Intergel, varying the degree of cross-linking and wetting, and goniometry, viscometry, and dynamic mechanical analysis were performed. It was found that increasing cross-linking could augment the contact angle, viscosity, storage modulus, and loss modulus, while wetting and lubrication were compromised. It is possible to create materials with mechanical properties softer than or comparable to those of abdominal organs that would be least disruptive to the body.

A cross-linked hyaluronate (HA)/collagen membrane was fabricated, and its efficacy in preventing peridural adhesion was evaluated in adult New Zealand white rabbits [86]. Membranes with different weight ratios of HA/collagen (60/40 and 40/60) were prepared by casting the solutions and drying. They were then immersed in 2.5% EDC solution in 80% acetone (pH = 4.7) at 35°C for 2 h to introduce cross-linking, followed by washing. The animals underwent L6 laminectomy and received the membrane on the exposed dura, while the control group received no treatment. Three months after surgery, the amount of scarring decreased significantly at laminectomy sites treated with either of the membranes. The extent of peridural adhesion decreased significantly in the 40/60 membrane, while the decrease was not significant in the 60/40 membrane.

In situ cross-linkable HA hydrogel was synthesized by chemical modification of HA with adipic dihydrazide (HA-ADH) and aldehyde (HA-CHO), and the efficacy of these materials in preventing postsurgical peritoneal adhesions was studied using a rabbit sidewall defect–cecum abrasion model [87]. Upon contact of the two HA derivatives, the gel formed in 3 s to 3 min, depending on the volumes of the two components. HA is degraded by endogenous hyaluronidase and by hydroxyl radicals. As shown in Fig. 5.22, the gel degradation kinetics could be varied by the precursor concentrations. For application to the site of injury in a rabbit model, the

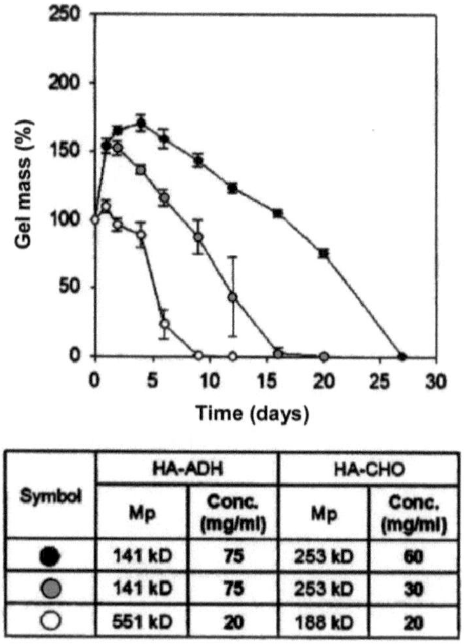

Fig. 5.22 Graph showing the effect of the gel concentration on degradation kinetics of macroscopic gel. Legends indicate the molecular weights (Mp) and concentrations of HA-ADH or HA-CHO in mg/mL. Values are expressed as mean ± standard deviation (n=4) [87]

Table 5.7 Evaluation of peritoneal adhesions [87]

	No treatment (n=12)		HAX (n=8)	
	Frequency	Percentage	Frequency	Percentage
Score 3	10	83	2	35
Score 2	0	0	1	12.5
Score 1	0	0	1	12.5
No adhesion	2	17	4	50
Median adhesion score	3 (3–3)		0.5 (0–2.25)	
% weight change	−5.7 ± 4.6		−9.0 ± 7.4	

HAX cross-linked hyaluronic acid

gel precursor solutions of 5 mL of HA-ADH (20 mg/mL) and 5 mL of HA-CHO (20 mg/mL) were placed in separate syringes, which were connected to a Baxter dual valve applicator, and coextruded through a 15-gauge needle. Table 5.7 shows the results of evaluation of peritoneal adhesions. Adhesion formation was significantly reduced in the HA gel group compared to that of the untreated control.

Ito et al. investigated the effect of in situ cross-linking for hydrogels of HA and cellulose derivatives [88]. Hybridization of HA with other biocompatible polysaccharides that are not degraded enzymatically in humans would slow degradation while preserving good biocompatibility. They prepared hydrogels by mixing hydrazide-modified hyaluronic acid (HA) with aldehyde-modified versions of cellulose derivatives such as CMC, hydroxypropylmethyl cellulose (HPMC), and methyl cellulose (MC) (Fig. 5.23). The physical properties of these gels are shown

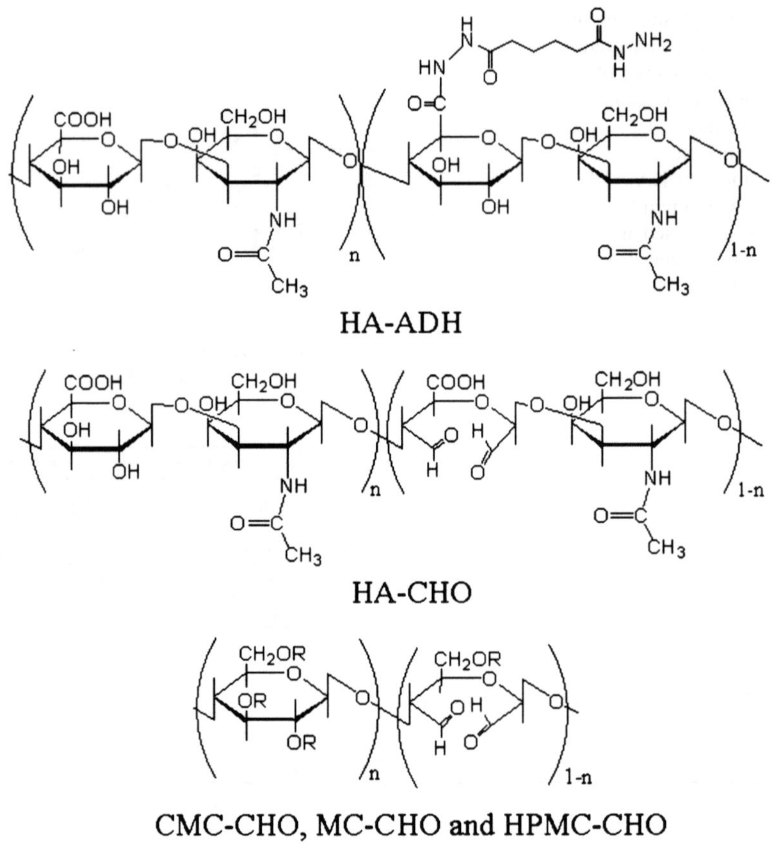

Fig. 5.23 Chemical structures of cellulose derivatives. CMC–CHO, R is H or CH_2COOH; HPMC–CHO, R is H or $(CH_2CH(CH_3)O)_nH$; MC-CHO, R is H or CH_3 [88]

Table 5.8 Physical properties of cross-linked hydrogels [88]

	Gelation time (s)	Swelling ratio[a] (%)	Modulus[a] (Pa)
HAX	3.5±1.0	206±14	32.4±17.2
HA-CMC	18.5±1.7	225±10	91.7±19.3
HA-HPMC	4.0±1.2	123±13	291.7±108.6
HA-MC	5.8±2.9	152±4	296.8±40.7

Data are average±SD (n=4)
HA hyaluronic acid, *CMC* carboxymethyl cellulose, *HPMC* hydroxypropylmethyl cellulose, *MC* methyl cellulose
[a]Measured on day 5 of immersion in phosphate-buffered saline

in Table 5.8. Gelation of these hydrogels took less than 1 min, and the gels had higher shear moduli than HA-HA gel. As shown in Table 5.9, all the cellulose-derived gels reduced the area of adhesion formation in a rabbit model of sidewall defect–bowel abrasion.

Table 5.9 Effectiveness of the hydrogels in preventing peritoneal adhesions in a rabbit model [88]

	HA–CMC	HA–HPMC	HA–MC	Control (saline)
Postoperative animal weight loss (%)	6.5 ± 3.6	2.8 ± 2.6	8.4 ± 3.3	11.7 ± 2.4
Score 3	1	2	0	3
Score 2	1	0	0	1
Score 1	0	0	1	0
No adhesion	2	2	3	0
Median adhesion score	1	2	0	3
Adhesion area (cm^2)	2.2 ± 3.3	0.3 ± 0.6	0.0 ± 0.0	13.1 ± 1.9

Animal weight loss refers to the weight loss in the week following the surgery. Adhesion area is the total area of adhesions with scores of 2 and 3. Weight change and adhesion area are expressed as average ± standard deviation ($n = 4$ per group)

Pharmacologically active compounds have been tested for prevention of postsurgical adhesions. However, administration of these agents has yielded ambiguous and potentially serious side effects. Combinations of these drugs with adhesion barriers, especially in gel forms, have been investigated for use in slow release systems. Ito and colleagues [89] synthesized an injectable hydrogel composed of cross-linkable HA conjugated to dexamethasone, a nonsteroidal anti-inflammatory drug. The synthesis of gel is shown in Fig. 5.24. The hydrogel was formed within 1 min by cross-linking reaction. The hydrogel was degraded in media over 5 days, and dexamethasone was released slowly over that time. The gels were subcutaneously injected in the dorsal midline of mice, and after 2 days, tissues were histologically assessed. The HA gel incorporating dexamethasone exhibited less inflammation than the unmodified cross-linked HA.

The in situ cross-linked HA hydrogel entrapping poly(glycolide-co-lactide) (PGLA) nanoparticles was investigated [90]. Intraperitoneal application of microparticles (5–250 μm) composed of PGLA caused a high rate of peritoneal adhesions in mice, whereas adhesion formation of PGLA nanoparticles (~265 nm) was less due to their rapid clearance from the peritoneum. HA hydrogel retained the nanoparticles in peritoneum while acting as a conventional barrier device, as well as preventing the retained polymer from causing adhesions. The hydrogel containing nanoparticles was formed in less than 10 s. The hybrid gel maintained the polymeric nanoparticles in the peritoneum for 1 week in mouse, and then cleared in less than 2 days. No animals developed adhesions. The incorporation of nanoparticles did not adversely affect in situ cross-linkability or mechanical properties of the HA gel. In a rabbit adhesion model, the animals treated with the hybrid gel had no adhesion in 62.5% of cases, whereas only 4.2% of untreated animals had no adhesions.

Tissue-type plasminogen activator (tPA) was delivered using a highly cross-linked in situ forming hyaluronan gel [91]. tPA was deficient the first 2–3 days postoperatively. The nontreated control group developed strong adhesions in 100% of animals, while adhesion formation was significantly reduced to 44% in the treated group.

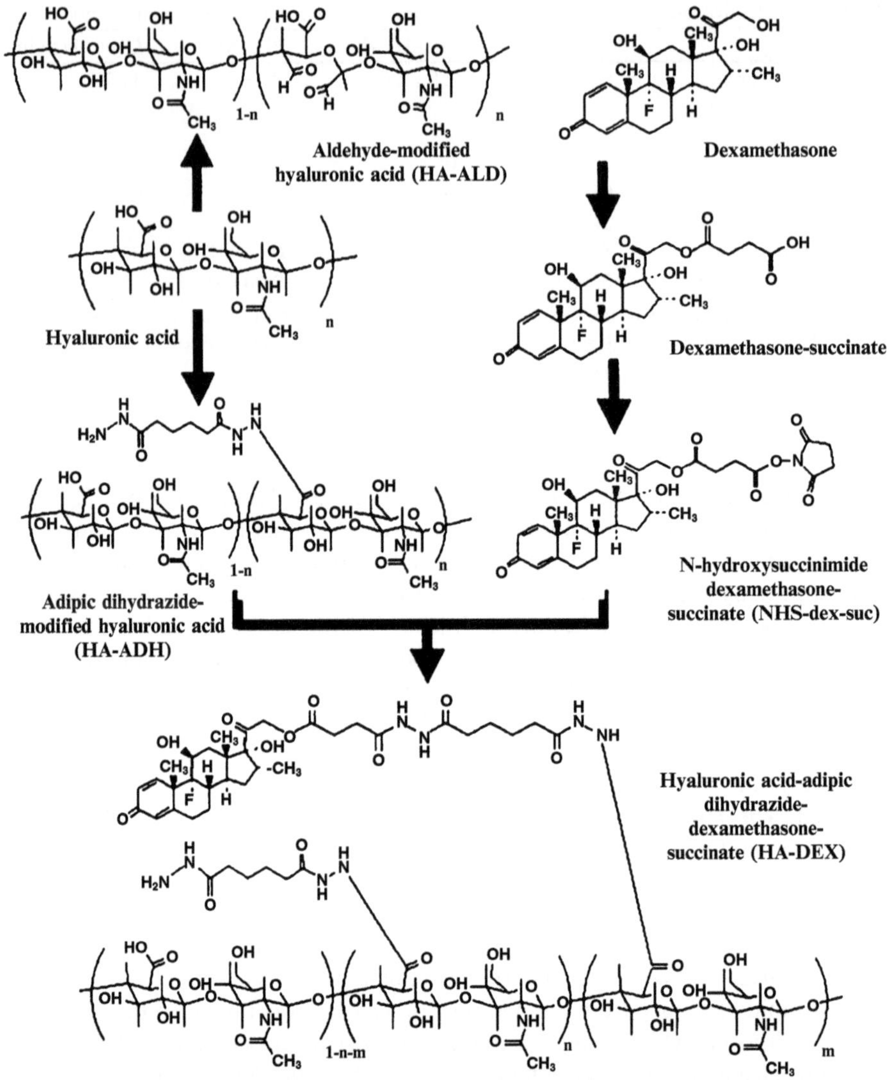

Fig. 5.24 Schematic representation of the mechanism of synthesis of cross-linked hyaluronic acid hydrogel containing dexamethasone (HAX-DEX). The final hydrogel is formed by mixing the aldehyde-derivatized hyaluronic acid with hyaluronic acid-adipic dihydrazide-dexamethasone succinate [89]

Synthetic peptides derived from human lactoferrin, PXL01, exhibits an inhibitory effect on the most important hallmarks of scar formation by reducing risk of infection, prohibiting inflammation, and promoting fibrinolysis [92]. The peptide was dissolved in sodium chloride solution and added to 2.5% HA solution to obtain 1.5–6 mg/mL PXL01 in 1.5% HA. In vitro release experiments revealed a burst release of PXL01 (approximately 70%) from the gel within 1 h. The efficacy of PXL01 in HA was evaluated using a sidewall defect–cecum abrasion model and a large bowel anastomosis healing model in rats. At 7 days after surgery, PXL01 with the HA gel had significantly reduced adhesion formation in rats.

Chitosan-Based Materials

Chitosan is another long-chain polysaccharide and is more cost-effective than HA. *N,O*-carboxymethyl chitosan (NOCC) is a derivative of chitosan. The addition of a carboxymethyl group to chitosan's nitrogen and oxygen centers produces a water-soluble, negatively charged polymer that is hydrophilic, lubricious, and viscoelastic. The efficacy of NOCC (2% NOCC solution and more viscous 1% cross-linked NOCC gel) in reducing adhesion formation was investigated using rat abdominal surgery models, including a uterine horn model and a small bowel laceration model [93]. After operation, 1 mL of 1% NOCC gel was applied directly to the site of the incision, and 3 mL of 2% NOCC solution was distributed evenly across the incision site before the closure. NOCC significantly reduced the incidence and intensity of postoperative adhesions in both the models. Wound healing at the site in neither of the models was deleteriously affected by the treatment with NOCC.

Gel was synthesized from succinyl chitosan and dextran aldehyde, and its efficacy in reducing adhesion formation in the intraperitoneal cavity was evaluated in rat models [94]. The surgical procedure was performed by laparotomy and either cecal abrasion or anastomotic simulation by enterotomy of the cecum with primary closure. After 21 days postoperatively, adhesions were significantly reduced for the treatment group in both the models. There was no adverse effect on wound healing.

The efficacy of in situ cross-linkable hydrogel made from *N*-carboxyethyl chitosan and oxidized dextran as an adhesion prevention barrier was evaluated in a rat cecal abrasion model and compared to that of Seprafilm [95]. Both Seprafilm and 2% hydrogel significantly decreased adhesion formation compared to the untreated control. The application of hydrogel was easier than that of Seprafilm, and the hydrogel adapted well on complex tissue geometries.

PEG-Based Materials

The efficacy of DuraSeal in pericardial adhesion prevention was evaluated with a porcine model [96]. After 6 weeks, all animals exhibited significant pericardial adhesions, and there was no significant difference in tenacity, extension, or strength of adhesions between the DuraSeal and nontreated sides of the hearts.

A biodegradable hydrogel based on CMC and PEG was prepared by the γ-irradiation cross-linking technique. Irradiation has been thought to be suitable for the formation of hydrogels since it does not require addition of any initiators or cross-linkers

Table 5.10 Evaluation of adhesions in rats treated with CMC/PEG hydrogel (Cp88/12) and with a solution of Cp88/12 [97]

Aggregation	Adhesion degree	Adhesion area (cm^2)	Adhesion strength
Controls	4.6±0.5	3.61±0.24	3.6±0.51
2% solution (Cp88/12)	0.2±0.42*	0.0125±0.04*	0.3±0.94*
Hydrogel (Cp88/12)	0.1±0.3*	0.0104±0.03*	0.2±0.63*

*$p<0.05$ versus controls

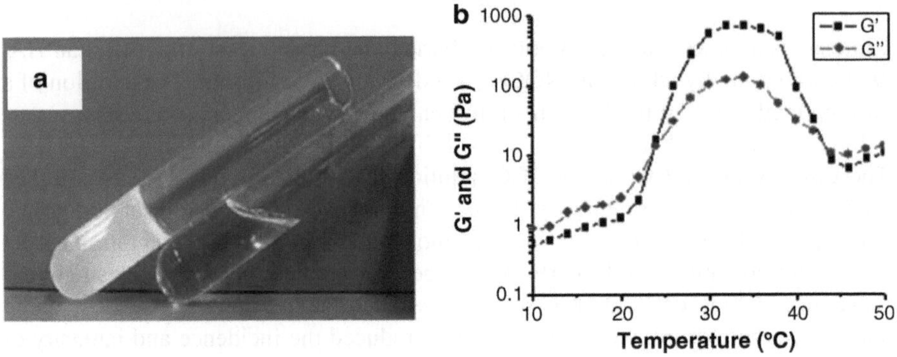

Fig. 5.25 (**a**) Photograph of the PECE hydrogel at 37°C (*left*) and 4°C (*right*). (**b**) Rheological analysis of the PECE hydrogel with respect to temperature. *G'* storage modulus, *G"* loss modulus [98]

that could be harmful and difficult to remove after gel formation. Polysaccharides such as CMC, carboxymethyl starch, carboxymethyl chitin, and carboxymethyl chitosan can be cross-linked by radiation in more than 10% aqueous solutions [97]. A mixture of CMC and PEG was exposed to 25 kGy of γ rays to prepare hydrogel. The efficacy of CMC/PEG hydrogel (with a dried content of CMC/PEG of 11.2% and a weight composition of 88/12) and its 2% solution was evaluated for adhesion prevention in rat peritoneal adhesion models. Two weeks after surgery, adhesion formation was significantly reduced in the hydrogel-treated groups compared to the untreated control group (Table 5.10).

Yang et al. investigated the efficacy of injectable poly(ethylene glycol)-poly(ε-caprolactone)-poly(ethylene glycol) (PEG-PCL-PEG) triblock copolymer hydrogel (PECE hydrogel) in preventing adhesions in a rat uterine horn model [98]. This hydrogel was found to be thermosensitive, biocompatible, and bioabsorbable. As shown in Fig. 5.25a, the hydrogel was a transparent flowing sol at 4°C, but became a gel at 37°C. No chemical reactions or cross-linkers were required for this system that is advantageous for the medical application. The reaction was reversible. Figure 5.25b shows the change in storage modulus (G') and loss modulus (G") of the hydrogel (25 wt.%) as a function of temperature. Gelation time was defined as the time point when G' becomes higher than G" at 24°C. An adhesion model was created by abrading the peritoneum of the abdominal wall and the uterine horns of rat. The PECE hydrogel was applied to the injured surfaces. During 12 days

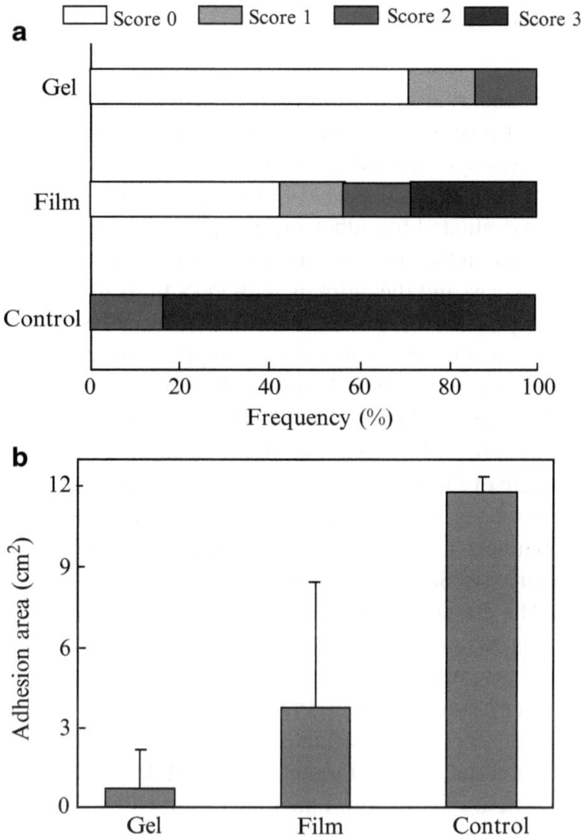

Fig. 5.26 (**a**) Distribution of adhesion scores of rabbits treated with the PCLA-PEG-PCLA hydrogel (group "gel"), the PLA film (group "film"), or without using any barrier materials (group "control") (score 0: no adhesion; score 1: mild, easily separable intestinal adhesion; score 2: moderate intestinal adhesion, separable by blunt dissection; score 3: severe intestinal adhesion, adhesion requiring sharp dissection). (**b**) Statistic results of area of intestinal adhesion to the defected abdominal wall (gel vs. film: $p=0.12$; gel vs. control: $p=0.00000$; film vs. control: $p=0.0015$). The initial abrasion damage covered 12 cm^2 [99]

postoperatively, none of the animals treated with the hydrogel ($n=12$) formed adhesions, whereas all the control animals ($n=12$) exhibited strong adhesions. The hydrogel disappeared in 7–9 days.

Zhang et al. synthesized a biodegradable triblock copolymer poly(ε-caprolactone-co-lactide)-*b*-poly(ethylene glycol)-*b*-poly(ε-caprolactone-co-lactide) (PCLA-PEG-PCLA) by ring-opening polymerization of CL and LA in the presence of PEG and stannous octoate as a catalyst [99]. This polymer solution, which is a sol at room temperature, rapidly formed a gel when warmed to the body temperature due to percolation of a self-assembled micelle network. The hydrogel implanted subcutaneously into the rabbit backs slowly degraded by hydrolysis and disappeared after 56 days. The adhesion prevention effect of this gel was investigated using a rabbit model of sidewall defect-bowel abrasion, and compared to that of a commercially available PLA film. As shown in Fig. 5.26, the hydrogel significantly reduced the adhesion formation compared to control (i.e., no treatment). Severe adhesion formation was observed for some animals treated with PLA film, but not for the hydrogel-treated group.

Other Materials

A bioabsorbable film of poly(glycerol sebacate) (PGS) was investigated for prevention of postoperative adhesion in a rat peritoneal adhesion model [100]. This polymer was synthesized by a polycondensation reaction between glycerol and sebacic acid, both of which have FDA approval for use as medical devices. The animals were evaluated for adhesions at 3, 5, and 8 weeks. Statistically significant reduction (94%) in adhesion formation rate was observed between the control animals (75% incidence) and the animals with PGS films (4.8%).

A powder form of antiadhesive material has been investigated, since the powder form has the advantage of easy application from ports in video-assisted operations compared with films. Izumi et al. evaluated the antiadhesive effect of cross-linked poly(gamma-glutamic acid) (BioPGA, Meiji Seika Ltd., Tokyo, Japan) in a rat model [101]. Poly(gamma-glutamic acid), a bioabsorbable polymer produced from a strain of *Bacillus subtilis*, can be cross-linked by gamma-irradiation. When applied in the body, it forms a viscous hydrogel by absorbing the surrounding fluid. The antiadhesive effect of cross-linked poly(gamma-glutamic acid) was found to be superior to Seprafilm® and Interceed®.

The effects of liquid adhesion barriers (Adept®, Hyalobarrier Gel, phospholipid emulsion) and solid barriers (poly-DL-lactide (PDLA) membrane and foil, Interceed®) were investigated in a rat uterine horn model and compared with that of lactated Ringer's solution [102]. PDLA membrane and foil were synthesized by the German Centre for Biomaterials and Organ Replacement (Deutsches Zentrum Für Biomaterialien und Organersatz; BMOZ). PDLA1 is a porous membrane with a thickness of 100 μm, and PDLA2 and PDLA3 are dense foils with thickness of 40–50 μm and 80–100 μm, respectively. Standard surgical injuries were created to the parietal and visceral peritoneum and the uterine horns. After 31 days, a second-look laparoscopy was performed for the barrier agents applied to assess adhesion formation. The results for adhesion severity and area coverage are shown in Fig. 5.27a, b, respectively. It was found that all barriers except PDLA3 were significantly more effective in preventing adhesion than lactated Ringer's solution. Adept® was found to be effective, easy to handle, and cost-effective.

Conclusion

Postsurgical adhesions have been a significant health problem with major implications for both quality of life and health-care costs. General intraoperative preventative practices such as use of starch-free gloves, avoidance of unnecessary peritoneal dissection, and avoidance of spillage of intestinal contents or gallstones may reduce the risk of adhesions, but the use of adhesion barriers is necessary in high-risk procedures. Many substances and materials have been used to decrease adhesion formation, and commercial products such as ePTFE, Interceed®, and Seprafilm have

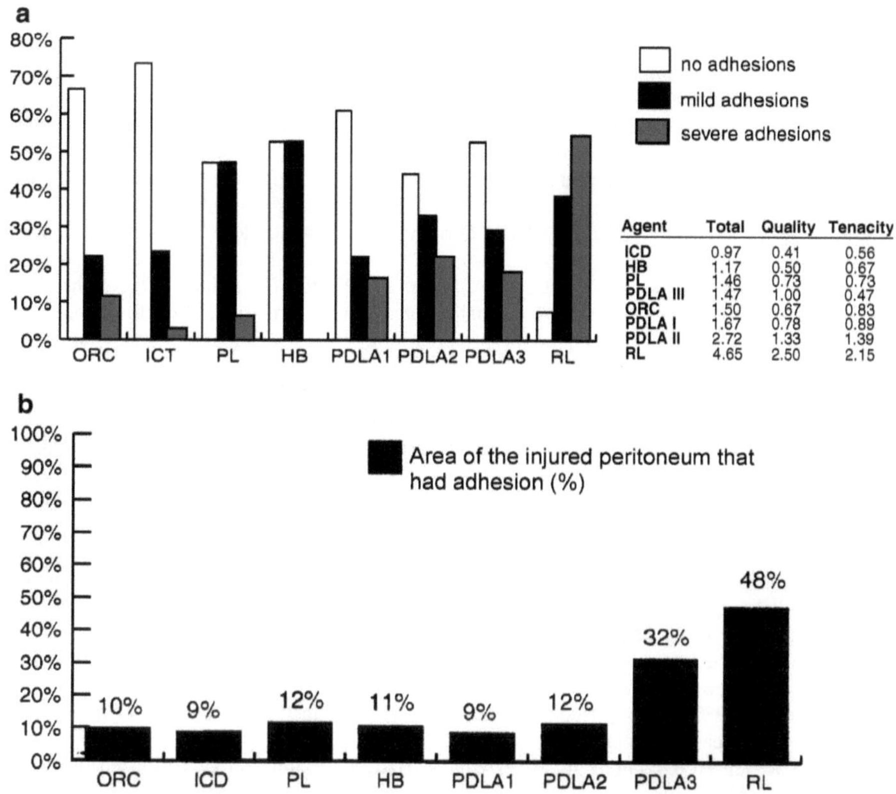

Agent	Total	Quality	Tenacity
ICD	0.97	0.41	0.56
HB	1.17	0.50	0.67
PL	1.46	0.73	0.73
PDLA III	1.47	1.00	0.47
ORC	1.50	0.67	0.83
PDLA I	1.67	0.78	0.89
PDLA II	2.72	1.33	1.39
RL	4.65	2.50	2.15

Fig. 5.27 Histograms showing the (**a**) adhesion severity and (**b**) adhesion area for the tested samples. *ORC* oxidized regenerated cellulose (Interceed®), *ICD* icodextrin (Adept®), *PL* phospholipid emulsion, *HB* hyaluronic acid (Hyalobarrier), *RL* lactated Ringer's solution [102]

been used. However, there is still no unequivocally effective adhesion-reducing barrier. Continuing research is providing advances in the field which promise the development of more effective products for reduction of adhesions.

References

1. Ray NF, Denton WG, Thamer M, Henderson SC, Perry S (1998) Abdominal adhesiolysis: inpatient care and expenditures in the United States in 1994. J Am Coll Surg 186(1):1–9
2. Imai A, Takagi H, Matsunami K, Suzuki N (2010) Non-barrier agents for postoperative adhesion prevention: clinical and preclinical aspects. Arch Gynecol Obstet 282(3):269–75
3. Tanaka A, Abe T, Matsuura A (2000) Prevention of postoperative intrapleural adhesion of the thoracotomy incision by a bioresorbable membrane in the rat adhesion model. Ann Thorac Cardiovasc Surg 6(3):151–60
4. Kamel RM (2010) Prevention of postoperative peritoneal adhesions. Eur J Obstet Gynecol Reprod Biol 150(2):111–8

5. Malinak LR, Young A (1997) Peritoneal closure: when and why. Contemp Obstet Gynecol 42:102–12
6. Tulandi T, Al-Jaroudi D (2003) Nonclosure of peritoneum: a reappraisal. Am J Obstet Gynecol 189(2):609–12
7. Tulandi T, Hum HS, Gelfand MM (1988) Closure of laparotomy incisions with or without peritoneal suturing and second-look laparoscopy. Am J Obstet Gynecol 158(3 Pt 1):536–7
8. Johns A (2001) Evidence-based prevention of post-operative adhesions. Hum Reprod Update 7(6):577–9
9. Cheong YC, Bajekal N, Li TC (2001) Peritoneal closure—to close or not to close. Hum Reprod 16(8):1548–52
10. Novitsky YW, Harrell AG, Cristiano JA, Paton BL, Norton HJ, Peindl RD, Kercher KW, Heniford BT (2007) Comparative evaluation of adhesion formation, strength of ingrowth, and textile properties of prosthetic meshes after long-term intra-abdominal implantation in a rabbit. J Surg Res 140(1):6–11
11. Harris ES, Morgan RF, Rodeheaver GT (1995) Analysis of the kinetics of peritoneal adhesion formation in the rat and evaluation of potential antiadhesive agents. Surgery 117:663–9
12. Buckman RF, Woods M, Sargent L, Gervin AS (1976) A unifying pathogenetic mechanism in the etiology of intraperitoneal adhesions. J Surg Res 20(1):1–5
13. Matsuda S, Se N, Iwata H, Ikada Y (2002) Evaluation of the antiadhesion potential of UV cross-linked gelatin films in a rat abdominal model. Biomaterials 23:2901–8
14. Larsson B, Lalos O, Marsk L, Tronstad SE, Bygdeman M, Pehrson S, Joelsson I (1985) Effect of intraperitoneal instillation of 32% dextran 70 on postoperative adhesion formation after tubal surgery. Acta Obstet Gynecol Scand 64(5):437–41
15. Weinans MJ, Kauer FM, Klompmaker IJ, Wijma J (1990) Transient liver function disturbances after the intraperitoneal use of 32% dextran 70 as adhesion prophylaxis in infertility surgery. Fertil Steril 53(1):159–61
16. Tulandi T (1987) Transient edema after intraperitoneal instillation of 32% dextran 70. A report of five cases. J Reprod Med 32(6):472–4
17. Sites CK, Jensen BA, Glock JL, Blackman JA, Badger GJ, Johnson JV, Brumsted JR (1997) Transvaginal ultrasonographic assessment of Hyskon or lactated Ringer's solution instillation after laparoscopy: randomized, controlled study. J Ultrasound Med 16(3):195–9
18. Holtz G, Baker E, Tsai C (1980) Effect of thirty-two per cent dextran 70 on peritoneal adhesion formation and re-formation after lysis. Fertil Steril 33(6):660–2
19. Rose B (1987) Safety of Hyskon for routine gynecologic surgery. A case report. J Reprod Med 32(2)
20. Ruiz JM, Neuwirth RF (1992) The incidence of complications associated with the use of Hyskon during hysteroscopy: experience in 1793 consecutive patients. J Gynecol Surg 8(4):219–24
21. Kossi J, Gronlund S, Uotila-Nieminen M, Crowe A, Knight A, Keranen U (2008) The effect of 4% icodextrin solution on adhesiolysis surgery time at the Hartmann's reversal: a pilot, multicentre, randomized control trial vs lactated Ringer's solution. Colorect Dis 11:168–72
22. Johns DB, Keyport GM, Hoehler F, Johns DB, Keyport GM, Hoehler F, Zerega GS (2001) Reduction of postsurgical adhesions with intergel® adhesion prevention solution: a multicenter study of safety and efficacy after conservative gynecologic surgery. Fertil Steril 76(3): 595–604
23. FDA 2003 Safety Alert INTERGEL adhesion prevention solution. Available from: http://www.fda.gov/Safety/MedWatch/SafetyInformation/SafetyAlertsforHumanMedicalProducts/ucm170010.htm
24. Rice VM, Shanti A, Moghissi KS, Leach RE (1993) A comparative evaluation of Poloxamer 407 and oxidized regenerated cellulose (Interceed [TC7]) to reduce postoperative adhesion formation in the rat uterine horn model. Fertil Steril 59(4):901–6
25. Weis C, Odermatt E (2006) A-part gel - an efficient adhesion prevention barrier. J Biomed Mater Res Part B: Appl Biomater 82B:174–82
26. Weis C, Odermatt EK, Kressler J, Unke Z, Wehner T, Freytag D (2004) Poly(vinyl alcohol) membrane for adhesion prevention. J Biomed Mater Res 70B:191–202

27. Janigen B, Weis C, Odermatt EK, Obermaier R, Hopt UT (2009) The new adhesion prophylaxis membrane A-Part from in vitro testing to first in vivo results. J Biomed Mater Res 89B:293–9
28. Young P, Johns A, Templeman C, Witz C, Webster B, Ferland R, Diamond MP, Block K, di Zerega G (2005) Reduction of postoperative adhesions after laparoscopic gynecological surgery with Oxiplex/AP Gel: a pilot study. Fertil Steril 84(5):1450–6
29. Dunn R, Lyman MD, Edelman PG, Campbell PK (2001) Evaluation of the SprayGel™ adhesion barrier in the rat cecum abrasion and rabbit uterine horn adhesion models. Fertil Steril 75(2):411–6
30. Trew G (2006) Postoperative adhesions and their prevention. Rev Gynaecol Perinat Pract 6(1):47–56
31. Ferland R, Mulani D, Campbell PK (2001) Evaluation of a sprayable polyethylene glycol adhesion barrier in a porcine efficacy model. Hum Reprod 16(12):2718–23
32. Mettler L, Audebert A, Lehmann-Willenbrock E, Jacobs VR, Schive K (2003) New adhesion prevention concept in gynecological surgery. JSLS 7(3):207–9
33. Mettler L, Audebert A, Lehmann-Willenbrock E, Schive K, Jacobs VR (2003) Prospective clinical trial of SprayGel as a barrier to adhesion formation: an interim analysis. J Am Assoc Gynecol Laparosc 10(3):339–44
34. Tjandra JJ, Chan MK (2008) A sprayable hydrogel adhesion barrier facilitates closure of defunctioning loop ileostomy: a randomized trial. Dis Colon Rectum 51(6):956–60
35. Ferland R, Campbell PK (2008) Evaluation of SprayShield adhesion barrier system in a porcine model of gynecological surgery Covidien 2008. Available from: http://www.surgipeer.com/sprayshield/publications.php
36. Angiotech presents positive AdhibitTM data at the 19th annual European Congress of Obstetrics and Gynecology: surgical adhesion scores were threefold less in patients treated with AdhibitTM. Vancouver: Angiotech Pharmaceuticals; 2006; Available from: http://www.bcbiotech.ca/News/Member_Press_Releaeses/pr04070602.asp.
37. Salminen JT, Mattila IP, Puntila JT, Sairanen HI (2010) Prevention of postoperative pericardial adhesions in children with hypoplastic left heart syndrome. Interact CardioVasc Thorac Surg 12:270–272
38. An expanded polytetrafluoroethylene barrier (Gore-Tex Surgical Membrane) reduces post-myomectomy adhesion formation. The Myomectomy Adhesion Multicenter Study Group (1995). Fertil Steril 3(3):491–3.
39. Haney AF, Hesla J, Hurst BS, Kettel LM, Murphy AA, Rock JA, Rowe G, Schlaff WD (1995) Expanded polytetrafluoroethylene (Gore-Tex Surgical Membrane) is superior to oxidized regenerated cellulose (Interceed TC7+) in preventing adhesions. Fertil Steril 63(5):1021–6
40. Hurst BS (1999) Permanent implantation of expanded polytetrafluoroethylene is safe for pelvic surgery. Human Reprod 14(4):925–7
41. Kawamura H, Yokota R, Yokota K, Watarai H, Tsunoda Y, Yamagami H, Hata T, Tanaka K, Masuko H, Ishizu H, Okada K, Adachi T, Kondo Y (2010) A sodium hyaluronate carboxymethylcellulose bioresorbable membrane prevents postoperative small-bowel adhesive obstruction after distal gastrectomy. Surg Today 40:223–7
42. Medina M, Paddock HN, Connolly RJ, Schwaitzberg SD (1995) Novel antiadhesion barrier does not prevent anastomotic healing in a rabbit model. J Invest Surg 8(3):179–86
43. Burns JW, Colt MJ, Burgees LS, Skinner KC (1997) Preclinical evaluation of Seprafilm bioresorbable membrane. Eur J Surg Suppl 577:40–8
44. Rajab TK, Wallwiener M, Planck C, Brochhausen C, Kraemer B, Wallwiener CW (2010) A direct comparison of seprafilm, adept, intercoat, and spraygel for adhesion prophylaxis. J Surg Res 161:246–9
45. Becker JM, Dayton MT, Fazio VW, Beck DE, Stryker SJ, Wexner SD, Wolff BG, Roberts PL, Smith LE, Sweeney SA, Moore M (1996) Prevention of postoperative abdominal adhesions by a sodium hyaluronate-based bioresorbable membrane: a prospective, randomized, double-blind multicenter study. J Am Coll Surg 183(4):297–306

46. Kusunoki M, Ikeuchi H, Yanagi H, Noda M, Tonouchi H, Mohri Y, Uchida K, Inoue Y, Kobayashi M, Miki C, Yamamura T (2005) Bioresorbable hyaluronate-carboxymethylcellulose membrane (Seprafilm) in surgery for rectal carcinoma: a prospective randomized clinical trial. Surg Today 35(11):940–5
47. Hayashi S, Takayama T, Masuda H, Kochi M, Ishii Y, Matsuda M, Yamagata M, Fujii M (2008) Bioresorbable membrane to reduce postoperative small bowel obstruction in patients with gastric cancer: a randomized clinical trial. Ann Surg 247(5):766–70
48. Beck DE, Cohen Z, Fleshman JW, Kaufman HS, van Goor H, Wolff BG, Adhesion Study Group Steering Committee (2003) A prospective, randomized, multicenter, controlled study of the safety of Seprafilm adhesion barrier in abdominopelvic surgery of the intestine. Dis Colon Rectum 46(10):1310–9
49. Fazio VW, Cohen Z, Fleshman JW, van Goor H, Bauer JJ, Wolff BG, Corman M, Beart RW Jr, Wexner SD, Becker JM, Monson JR, Kaufman HS, Beck DE, Bailey HR, Ludwig KA, Stamos MJ, Darzi A, Bleday R, Dorazio R, Madoff RD, Smith LE, Gearhart S, Lillemoe K, Gohl J (2006) Reduction in adhesive small-bowel obstruction by Seprafilm adhesion barrier after intestinal resection. Dis Colon Rectum 49(1):1–11
50. Walther T, Rastan A, Dähnert I, Falk V, Jacobs S, Mohr FW, Kostelka M (2005) A novel adhesion barrier facilitates reoperations in complex congenital cardiac surgery. J Thorac Cardiovasc Surg 129(2):359–63
51. Naito Y, Shin'oka T, Hibino N, Matsumura G, Kurosawa H (2008) A novel method to reduce pericardial adhesion: a combination technique with hyaluronic acid biocompatible membrane. J Thorac Cardiovasc Surg 135(4)):850–6
52. Wiseman DM, Trout JR, Franklin RR, Diamond MP (1999) Metaanalysis of the safety and efficacy of an adhesion barrier (Interceed TC7) in laparotomy. J Reprod Med 44(4):325–31
53. Prevention of postsurgical adhesions by INTERCEED(TC7), an absorbable adhesion barrier: a prospective randomized multicenter clinical study. INTERCEED(TC7) Adhesion Barrier Study Group (1989). Fertil Steril 51(6):933–8
54. Iliopoulos J, Cornwall GB, Evans RO, Manganas C, Thomas KA, Newman DC, Walsh WR (2004) Evaluation of a bioabsorable polylactide film in a large animal model for the reduction of retrosternal adhesions. J Surg Res 118(2):144–53
55. Faerden A, Reiertsen O, Marx A (2006) The effectiveness of a bioresorbable polymer sheet (SurgiWrap®) in minimizing Soft Tissue Attachments (STA) after major colorectal surgery: a randomized controlled pilot study. MAST Biosurgery from manufactures website.
56. Okuyama N, Wang CY, Rose EA, Rodgers KE, Pines E, Zerega GS, Oz MC (1999) Reduction of retrosternal and pericardial adhesions with rapidly resorbable polymer films. Ann Thorac Surg 68((3):913–8
57. Lodge AJ, Wells WJ, Backer CL, O'Brien JE, Austin EH, Bacha EA, Decampli WM, Lavin PT, Weinstein S (2008) A novel bioresorbable film reduces postoperative adhesions after infant cardiac surgery. Ann Thorac Surg 86(2):614–21
58. Gagnieu CH, Forest PO (2007) In vivo biodegradability and biocompatibility of porcine type I atelocollagen newly crosslinked by oxidized glycogen. Biomed Mater Eng 17(1):9–18
59. Bel A, Kachatryan L, Bruneval P, Peyrard S, Gagnieu C, Fabiani JN, Menasché P (2010) A new absorbable collagen membrane to reduce adhesions in cardiac surgery. Interact Cardiovasc Thorac Surg 10(2):213–6
60. Gruber-Blum S, Petter-Puchner AH, Brand J, Fortelny RH, Walder N, Oehlinger W, Koenig F, Redl H (2011) Comparison of three separate antiadhesive barriers for intraperitoneal onlay mesh hernia repair in an experimental model. Br J Surg 98(3):442–9
61. Diamond MP, Kruger M, Saed GM (2004) Effect of Tisseel on expression of tissue plasminogen activator and plasminogen activator inhibitor-1. Fertil Steril 81:1657–64
62. Lindenberg S, Lauritsen JG (1984) Prevention of peritoneal adhesion formation by fibrin sealant. Ann Chirurgiae et Gynaecologiae 73:11–3
63. Prieto-Díaz-Chávez E, Medina-Chávez JL, Ramírez-Barba EJ, Trujillo-Hernández B, Millán-Guerrero RO, Vásquez C (2008) Reduction of peritoneal adhesion to polypropylene mesh with the application of fibrin glue. Acta chir Belg 108:433–7

64. Martín-Cartes JA, Morales-Conde S, Suárez-Grau JM, Bustos-Jiménez M, Cadet-Dussort JM, López-Bernal F, Morcillo-Azcárate J, Tutosaus-Gómez JD, Morales-Méndez S (2008) Role of fibrin glue in the prevention of peritoneal adhesions in ventral hernia repair. Surg Today 38(2):135–40
65. Petter-Puchner AH, Walder N, Redl H, Schwab R, Ohlinger W, Gruber-Blum S, Fortelny RH (2008) Fibrin sealant (Tissucol) enhances tissue integration of condensed polytetrafluoroethylene meshes and reduces early adhesion formation in experimental intraabdominal peritoneal onlay mesh repair. J Surg Res 150(2):190–5
66. Frykman E, Jacobsson S, Widenfalk B (1993) Fibrin sealant in prevention of flexor tendon adhesions: an experimental study in the rabbit. J Hand Surg 18A:68–75
67. Tulandi T, Al-Jaroudi D (2007) Averting adhesions: surgical techniques and tools. OBG Management 19(05):86–94
68. Suzuki S, Ikada Y (in press) Sealing effects of cross-linked gelatin. J Biomater Appl
69. Yoshioka I, Saiki Y, Sakuma K, Iguchi A, Moriya T, Ikada Y, Tabayashi K (2007) Bioabsorbable gelatin sheets latticed with polyglycolic acid can eliminate pericardial adhesion. Ann Thorac Surg 84:864–870
70. Sakuma K, Iguchi A, Ikada Y, Tabayashi K (2005) Closure of the pericardium using synthetic bioabsorbable polymers. Ann Thorac Surg 80:1835–40
71. Young RL, Cota J, Zund G, Mason BA, Wheeler JM (1991) The use of an amniotic membrane graft to prevent postoperative adhesions. Fertil Steril 55(3):624–8
72. Miyata T, Fususe M, Yamane Y, Noishiki Y (1988) A biodegradable antiadhesion collagen membrane with slow release heparin. ASAIO Trans 34(3):687–91
73. Badawy SZ, Baggish MS, ElBakry MM, Baltoyannis P (1989) Evaluation of tissue healing and adhesion formation after an intraabdominal amniotic membrane graft in the rat. J Reprod Med 34((3):198–202
74. Edwards GA, Glattauer V, Nash TJ, White JF, Brock KA, Werkmeister JA, Ramshaw JA (1997) In vivo evaluation of a collagenous membrane as an absorbable adhesion barrier. J Biomed Mater Res 34(3):291–7
75. Arnold PB, Green CW, Foresman PA, Rodeheaver GT (2000) Evaluation of resorbable barriers for preventing surgical adhesions. Fertil Steril 73(1):157–61
76. Liu S, Boutrand JP, Tadie M (2001) Use of a collagen-based sealant to prevent in vivo epidural adhesions in an adult rat laminectomy model. J Neurosurg 94(1 Suppl)):61–7
77. Liu S, Boutrand JP, Bittoun J, Tadie M (2002) A collagen-based sealant to prevent in vivo reformation of epidural scar adhesions in an adult rat laminectomy model. J Neurosurg 97(1 Suppl)):69–74
78. Longaker MT, Chiu ES, Adzick NS, Stern M, Harrison MR, Stern R (1991) Studies in fetal wound healing. V. A prolonged presence of hyaluronic acid characterizes fetal wound fluid. Ann Surg 213(4):292–6
79. Mast BA, Diegelmann RF, Krummel TM, Cohen IK (1992) Scarless wound healing in the mammalian fetus. Surg Gynecol Obstet 174(5):441–51
80. Huang-Lee LL, Wu JH, Nimni ME (1994) Effects of hyaluronan on collagen fibrillar matrix contraction by fibroblasts. J Biomed Mater Res 28(1):123–32
81. Mensitieri M, Ambrosio L, Nicolais L, Bellini D, O'Regan M (1996) Viscoelastic properties modulation of a novel autocrosslinked hyaluronic acid polymer. J Mater Sci Mater Med 7(11):695–8
82. Renier D, Bellato P, Bellini D, Pavesio A, Pressato D, Borrione A (2005) Pharmacokinetic behaviour of ACP gel, an autocrosslinked hyaluronan derivative, after intraperitoneal administration. Biomaterials 26(26):5368–74
83. Belluco C, Meggiolaro F, Pressato D, Pavesio A, Bigon E, Donà M, Forlin M, Nitti D, Lise M (2001) Prevention of postsurgical adhesions with an autocrosslinked hyaluronan derivative gel. J Surg Res 100(2):217–21
84. De Iaco PA, Muzzupapa G, Bigon E, Pressato D, Donà M, Pavesio A, Bovicelli L (2001) Efficacy of a hyaluronan derivative gel in postsurgical adhesion prevention in the presence of inadequate hemostasis. Surgery 130(1):60–4

85. Vorvolakos K, Isayeva IS, Luu H-MD, Patwardhan DV, Pollack SK (2010) Ionically cross-linked hyaluronic acid: wetting, lubrication, and viscoelasticity of a modified adhesion barrier gel. Med Dev Evid Res 2011((4):1–10
86. Shih HN, Fang JF, Chen JH, Yang CL, Chen YH, Sung TH, Shih LY (2004) Reduction in experimental peridural adhesion with the use of a crosslinked hyaluronate/collagen membrane. J Biomed Mater Res B Appl Biomater 71(2):421–8
87. Yeo Y, Highley CB, Bellas E, Ito T, Marini R, Langer R, Kohane DS (2006) In situ cross-linkable hyaluronic acid hydrogels prevent post-operative abdominal adhesions in a rabbit model. Biomaterials 27(27):4698–705
88. Ito T, Yeo Y, Highley CB, Bellas E, Benitez CA, Kohane DS (2007) The prevention of peritoneal adhesions by in-situ cross-linking hydrogels of hyaluronic acid and cellulose derivatives. Biomaterials 28(6):975–83
89. Ito T, Fraser IP, Yeo Y, Highley CB, Bellas E, Kohane DS (2007) Anti-inflammatory function of an in situ cross-linkable conjugate hydrogel of hyaluronic acid and dexamethasone. Biomaterials 28(10):1778–86
90. Yeo Y, Ito T, Bellas E, Highley CB, Marini R, Kohane DS (2007) In situ cross-linkable hyaluronan hydrogels containing polymeric nanoparticles for preventing postsurgical adhesions. Ann Surg 245(5):819–24
91. Yeo Y, Bellas E, Highley CB, Langer R, Kohane DS (2007) Peritoneal adhesion prevention with an in situ cross-linkable hyaluronan gel containing tissue-type plasminogen activator in a rabbit repeated-injury model. Biomaterials 28(25):3704–13
92. Nilsson E, Björn C, Sjöstrand V, Lindgren K, Münnich M, Mattsby-Baltzer I, Ivarsson ML, Olmarker K, Mahlapuu M (2009) A novel polypeptide derived from human lactoferrin in sodium hyaluronate prevents postsurgical adhesion formation in the rat. Ann Surg 250(6):1021–8
93. Kennedy R, Costain DJ, McAlister VC, Lee TD (1996) Prevention of experimental postoperative peritoneal adhesions by N,O-carboxymethyl chitosan. Surgery 120(5):866–70
94. Lauder CI, Garcea G, Strickland A, Maddern GJ (2011) Use of a modified chitosan-dextran gel to prevent peritoneal adhesions in a rat model. J Surg Res 171(2):877–82
95. Falabella CA, Melendez MM, Weng L, Chen W (2010) Novel macromolecular crosslinking hydrogel to reduce intra-abdominal adhesions. J Surg Res 159(2):772–8
96. Pätilä T, Jokinen JJ, Salminen J, Kankuri E, Harjula A (2008) Polyglycolic acid glue does not prevent intrapericardial adhesions in a short-term follow-up. J Surg Res 148((2):181–4
97. Lee J-H, Nho Y-C, Lim Y-M, Son T-I (2005) Prevention of surgical adhesions with barriers of carboxymethylcellulose and poly(ethylene glycol) hydrogels synthesized by irradiation. J Appl Polym Sci 96(4):1138–45
98. Yang B, Gong CY, Qian ZY, Zhao X, Li ZY, Zhou ST, Qi XR, Zhong Q, Luo F, Wei YQ (2011) Prevention of abdominal adhesion formation by thermosensitive PECE-hydrogel in a rat uterine horn model. J Biomed Mater Res B Appl Biomater 96(1):57–66
99. Zhang Z, Ni J, Chen L, Yu L, Xu J, Ding J (2011) Biodegradable and thermoreversible PCLA-PEG-PCLA hydrogel as a barrier for prevention of post-operative adhesion. Biomaterials 32(11):4725–36
100. Pryor HI, O'Doherty E, Hart A, Owens G, Hoganson D, Vacanti JP, Masiakos PT, Sundback CA (2009) Poly(glycerol sebacate) films prevent postoperative adhesions and allow laparoscopic placement. Surgery 146:490–7
101. Izumi Y, Yamamoto M, Kawamura M, Adachi T, Kobayashi K (2006) Cross-linked poly(gamma-glutamic acid) attenuates peritoneal adhesion in a rat model. Surgery 141(5):678–81
102. Wallwiener M, Brucker S, Hierlemann H, Brochhausen C, Solomayer E, Wallwiener C (2006) Innovative barriers for peritoneal adhesion prevention: liquid or solid? A rat uterine horn model. Fertil Steril 86(4 Suppl):1266–76

Chapter 6
Devices for Bone Fixation

Abstract Bone fixation devices made of bioabsorbable polymers have advantages over traditional metallic implants as the latter require a secondary operation to remove the device. However, bone fixation devices are expected to have much higher mechanical strengths than polymers used in soft tissue applications. Poly(L-lactide) (PLLA) is a semicrystalline bioabsorbable polymer that can be drawn to increase its mechanical strength. The present chapter describes the chemistry of PLLA and its applications in bone fixation devices. Degradation of PLLA takes a few years which may be adequate for bone regeneration. An important issue of bioabsorbable fixation devices is the balance between the bioabsorption rate and the mechanical strength maintenance. It is known that PLLA exhibits a piezoelectric effect that can accelerate new bone formation. Currently, PLLA bone fixation devices, including screws, pins, rods, and plates, are clinically available. They are used preferentially in non-load-bearing applications such as in maxillofacial surgery.

Introduction

One operation widely performed in orthopedic, oral, and plastic surgeries is the repair of fractured bones. As the number of elderly people in society increases, so does the overall number of bone fractures. This increase in the incidence of fractures is especially attributable to the incidence of falling in elderly people living in residential areas. Fractured bones are primarily repaired using fixation devices such as wires, nails, pins, screws, and plates. Fixation devices are usually made of metals and, except for special cases, need to be removed in a secondary operation after full reunion of the fractured bones. Bioabsorbable fixation devices were introduced in the 1970s to eliminate the need for this secondary surgery.

The most critical property of bioabsorbable fixation devices is mechanical strength. The mechanical strength of these devices must, at a minimum, be higher than that of the bone. Many metals meet this requirement, but only a few plastics do. In addition, the strong plastics used in the construction of fixation devices must

S. Suzuki and Y. Ikada, *Biomaterials for Surgical Operation*,
DOI 10.1007/978-1-61779-570-1_6, © Springer Science+Business Media, LLC 2012

support the body load until bone reunion is complete, but should then be bioabsorbed. The achievement of a long-term balance between mechanical strength and material bioabsorption is a great challenge to biomaterials researchers. In 1971, Cutright et al. reported for the first time that poly(L-lactic acid) could be used for orthopedic fixation [1]. Since then, extensive studies have been conducted, and there are now a variety of commercially available bioabsorbable fixation devices, mostly fabricated from poly(lactic acid) (PLA). PLA fixation devices have been in use longer than any other bioabsorbable medical device, with the exception of synthetic bioabsorbable sutures. The chemistry of PLA will, therefore, be described first.

Chemistry of PLA

Because the molecular weight of PLA synthesized by polycondensation of lactic acids is too low for medical applications (<20,000), ring-opening polymerization of lactides has been used to obtain high-molecular-weight PLA. Poly(L-lactide or L-lactic acid) (PLLA) is a semicrystalline material, and its full degradation takes more than 2 years, whereas poly(D,L-lactide or D,L-lactic acid) (PDLLA) degrades within 16 months, but is amorphous. The toughness of PLLA can be enhanced by a drawing process. Drawing PLLA that has a molecular weight over 70,000 increases its chain orientation and crystallinity, as shown in Fig. 3.4. Drawing PLLA also increases its piezoelectricity, which will be discussed later. Another method for strengthening PLA is self-reinforcement (SR). In this technique, polymeric fibers with a high modulus are bound to a matrix of the same polymer without the use of adhesion promoters.

The degradation of PLA by hydrolytic scission of ester linkages yields lactic acid, which is a natural product associated with muscular construction in animals and humans. This acid can be decomposed by the body's normal metabolic pathways to pyruvic acid, which enters the tricarboxylic acid cycle to yield carbon dioxide and water. A study with carbon-labeled PLA revealed no significant accumulation of degradation products in any organs and found only very small amounts in feces and urine, thereby indicating that the products were released through respiration [2]. Because L-lactic acid is the naturally occurring stereoisomer of lactic acid, PLLA is more commonly used for medical applications than poly(D-lactic acid) (PDLA), which yields D-lactic acid when it degrades.

PLA, other poly(α-hydroxyacid)s, and their copolymers have been approved for implantation in the human body by the US FDA and other regulatory agencies in many countries. A number of these products are now commercially available and have been successfully used in the medical field. However, despite these successes, there are some disadvantages of PLA to be overcome, including its slow degradation and hydrophobic nature.

While the degradation product of PLA, lactic acid, can be decomposed by the body, this acid is relatively strong. The accumulation of lactic acid at the implant site because of a "burst" release from bulk degradation of PLA will lead to lowering of local pH, triggering an inflammation response. Inflammation lasting over 1 year has been observed in some cases [3]. Additionally, particles smaller than 2 μm are released during PLA degradation, and these particles can cause a foreign-body reaction that results in detrimental effects on the bone tissue [4]. There have been attempts to neutralize the acidic degradation products by adding agents such as calcium carbonates and/or calcium phosphates to the PLLA implants [5], but these effects are unclear.

In addition to lactic acid, the polymer products contain trace amounts of initiators, coinitiators, and catalysts. The most effective and widely used initiators for ring-opening polymerization of lactide are based on tin, typically tin(II) 2-ethylhexanoate. Due to its low toxicity, tin(II) 2-ethylhexanoate has been approved for this use by the FDA. The residual concentration of tin(II) 2-ethylhexanoate in the polymer is strictly controlled to a safe level. Many studies have been conducted to develop a new range of initiators and catalysts based on metals that are more biocompatible, such as magnesium and calcium.

The hydrophobic nature of PLA and other poly(α-hydroxyacid)s results in a low affinity to cells and proteins, suppressing tissue formation. The use of PLA as a scaffold for tissue engineering is limited due to this low tissue affinity and very slow degradation of the polymer.

Bone Fixation Applications

Biodegradable materials are used as substitutes for metallic devices in fractured bone fixation because metallic fixations have the following disadvantages (1) Secondary surgery is required to remove these materials after the reunion of fractured bones. The temporary weakening of the bone as a result of the removal of implants may lead to bone refracture. (2) Very stiff metallic materials may cause osteoporosis beneath the implant because of stress shielding and may lead to bone fracture at the implant site. (3) While relatively rare, corrosion of metallic implants is not a negligible concern. (4) Metallic implants interfere with or distort postoperative radiography, CT, and MRI images. (5) During postoperative radiotherapy in patients who have had tumors removed, the presence of metallic implants may alter the local dose distribution (i.e., overdosage proximal to the implant and underdosage distal to the implant).

The difficulties associated with metals and the advantages of biodegradable materials make degradable materials ideal for bone fixation. Because materials for this purpose require high strength, similar to that of bone, PLA is almost exclusively used in this field. Pietrzak extensively reviewed biodegradable fixation devices [6]. Table 6.1 summarizes the commercially available PLA bone fixation devices in the global market. Next, the effects of stress shielding and piezoelectricity of PLA will be described, followed by its application to bone fixation.

Table 6.1 Commercially available, bioabsorbable bone fixation devices

Manufacturer	Country	Products	Trade name	Materials
Stryker	United States	Plate	DeltaSystem	Composition of poly-L-lactide/D-lactide/glycolide with monomer ratio of 85/5/10
Biomet	United States	Plate, screw	LactoSorb®	Copolymer, 82% L-lactic acid and 18% glycolic acid
Bionx	United States	Pin	SmartPins	Self-reinforced poly(L-lactide)
ConMed Linvatec	United States	Screw	SmartScrews	Self-reinforced 96 L/4D copolymer
Zimmer Centerpulse	United States	Suture anchor	Resorbable Suture Anchor	PLLA
Tornier Inc.	United States	Suture anchor	Insite	Poly(L-lactide)
		Screw	RFS	L-lactic/co-glycolic acid copolymer (PLGA 85 L/15 G)
KLS Martin	United States	Pins	SonicWeld Rx	Homogeneous blend of D and L isomers of poly(lactic acid)
Arthrex	United States	Suture anchor	Push Lock	PLLA or PLDLA combined with TCP
J and J DePuy	United States	Pin	OrthoSorb	Poly(p-dioxanone) (PDS)
		Suture anchor	RAPIDLOC	PDS/PLA
Surgical Dynamics	United States	Suture anchor	SD sorb	82% PLLA, 18% PGA
Smith and Nephew (orthopedics)	United Kingdom	ACL screw	BioRCI, Endo-Fix	Polyglyconate
Phusis	France	Screw	Tenoscrews	PLA98
		Screw	Phusiline	PLA98
Synthes	Switzerland	Plate, foil, mesh, screw, tack	RapidSorb	Poly(L-lactide-co-glycolide)
Inion	Finland	Plate, mesh, screw	CPS	Degradable copolymers composed of L-lactic acid, D-lactic acid and trimethylene carbonate
Takiron	Japan	Screw, pin, washer	SuperFIXSORB 30	Composites of particulate hydroxyapatite (HA) and poly(L-lactide) (PLLA), HA/PLLA = 30/70(wt.%)
		Screw, pin, washer	OSTEOTRANS	Composites of particulate hydroxyapatite (HA) and poly(L-lactide) (PLLA), HA/PLLA = 30/70(wt.%)
Gunze	Japan	Screw, pin, plate	NEOFIX, GRAND FIX	Poly(L-lactic acid) (PLLA)

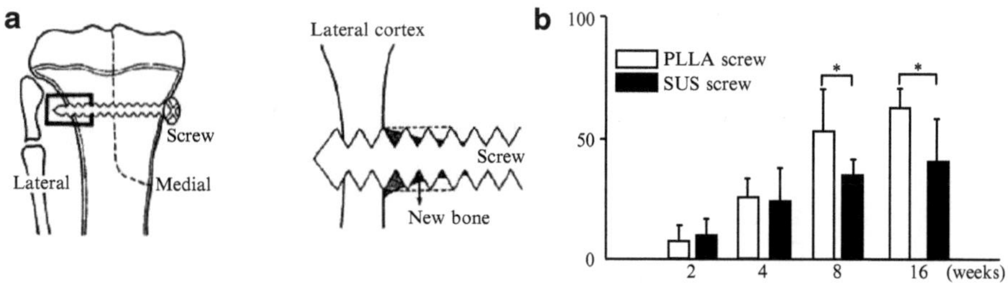

Fig. 6.1 (**a**) Diagram of screw implantation in rabbit tibia proximal osteotomy and new bone formation. (**b**) New bone formation around the PLLA and SUS screws expressed as percentage of the area of new bone relative to the indentations of screw threads [7]

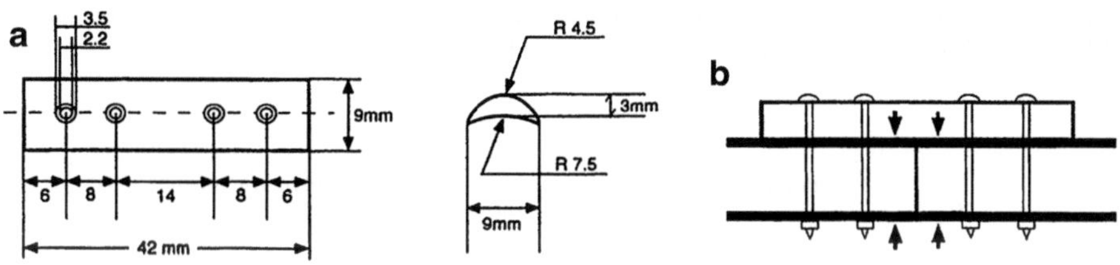

Fig. 6.2 (**a**) Design of PLLA and stainless steel plates. (**b**) Points where cortical thickness was measured beneath (*downward arrow*) and opposite (*upward arrow*) the plates [8]

Stress-Shielding Effect

Stress shielding is a mechanical concept. When two or more components with different elastic moduli make up a mechanical system, load, stress, and strain are redistributed to the stiffer component. This redistribution occurs between bones and metal fixation devices. In the early stages of healing, the stiffness of fixation devices is essential for the firm stabilization of fracture ends. However, the decrease in bone stress stimulation will lead to osteoporosis, especially in the later stages of healing. This osteoporosis can lead to new bone fractures and to the loosening of implants. Because biodegradable materials are slowly bioabsorbed, the transfer of load to the bone increases gradually, decreasing the stress-shielding effect during bone remodeling.

Stress-shielding effects of PLLA and stainless steel (SUS304) screws were investigated using a rabbit model under load-bearing conditions as shown in Fig. 6.1. An increase in new bone formation was observed with both types of screws, but a significantly larger amount of new bone was observed with PLLA screws than with stainless steel screws at 8 and 12 weeks after implantation [7].

In an examination of bone strength after the implantation of fixation devices, PLLA and stainless steel plates (design shown in Fig. 6.2) were implanted into

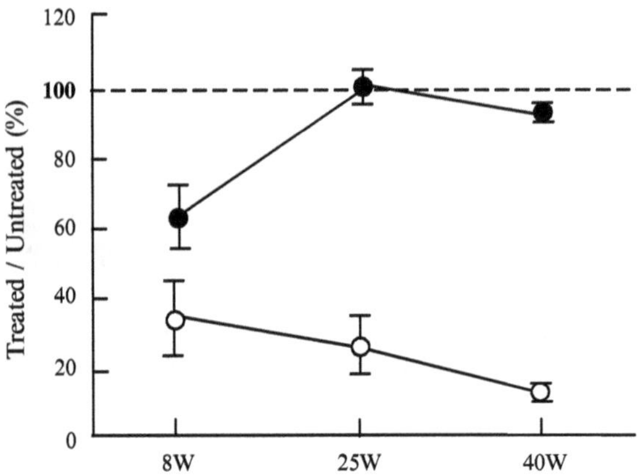

Fig. 6.3 Changes in the mechanical strength of united specimens fixed with three-point bending PLLA (*filled circle*) and stainless steel (*open circle*) [8]

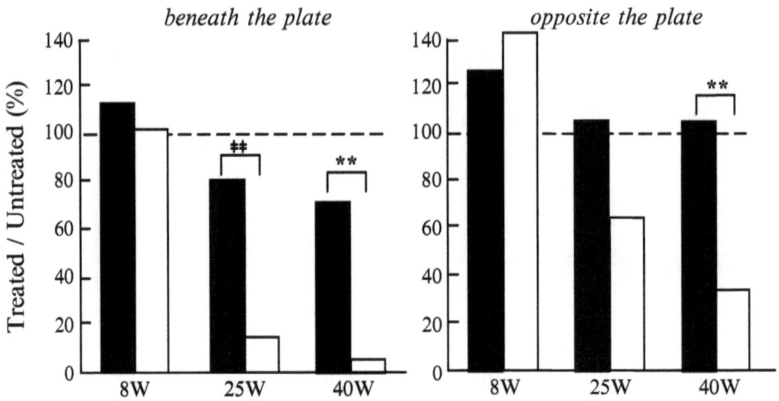

Fig. 6.4 Cortical thickness beneath (*left*) and opposite (right) PLLA (*filled square*) and stainless steel (*open square*) plates. Ordinate represents the treated/untreated ratio for each femur pair, expressed as a percentage [8]

rabbit femoral shaft osteotomies [8]. Figure 6.3 represents the mechanical strength of united specimens after 8, 25, and 40 weeks of implantation. Full femur strength was recovered after 25 weeks with PLLA plates, whereas the bones fixed with stainless steel plates showed only 30% of their original strength after 25 weeks and even less after 40 weeks. As shown in Fig. 6.4, stainless steel plates caused a decrease in bone thickness not only beneath but also above the plates. The result clearly demonstrates the stress-shielding effect of metal plates on bone strength. The strength of PLLA plates was reduced to almost zero after 25 weeks, eliminating the stress-shielding effect.

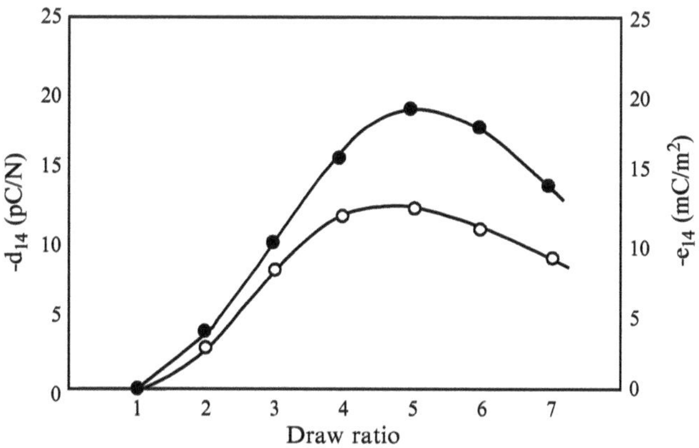

Fig. 6.5 The piezoelectric constants [stress: e_{14} (*filled circle*) and strain: d_{14} (*open circle*)] of PLLA film at room temperature as a function of draw ratio [10]

Piezoelectric Effect

It is well known that the bone growth is enhanced by the application of electrical stimulation [9]. Callus formation, which generally takes place at the beginning of bone formation, is promoted by applied electric fields. Some materials are capable of generating an electric potential when mechanical stress is applied. This property is called piezoelectricity. Most piezoelectric materials are crystals and some, such as bone, are ceramics. Poly(vinylidene fluoride) (PVDF) is a well-known piezoelectric material.

Interestingly, when a drawing process is applied to PLLA, it can exhibit a degree of piezoelectricity similar to that of PVDF, as shown in Fig. 6.5. PLLA obtains high piezoelectricity when its polymer chains are highly oriented by the drawing process because it is a crystalline polymer with an asymmetric carbon in the repeating unit of the main chain. On the other hand, PDLLA is amorphous and does not exhibit piezoelectric effects under any conditions. The electric potential induced in PLLA by additional stress promotes pseudo-bone formation.

The piezoelectric effect of drawn PLLA was investigated using films and rods [10]. As shown in Fig. 6.5, the piezoelectric stress (e_{14}) and strain (d_{14}) constants of PLLA film increased with drawing ratio up to a drawing ratio of 5, reaching maximum values of -20 mC/m$_2$ and -10 pC/N, respectively, which are comparable to those of PVDF. The increase in piezoelectric stress and strain constants is due to the increased alignment of the PLLA chains caused by drawing. Drawing ratios over 5 resulted in decreasing piezoelectric constants, possibly due to fibrilization, which causes disorder and discontinuity in the microscopic crystal arrangement.

To examine the effect of piezoelectricity on the healing of bone fractures, PLLA rods that differed only in piezoelectric constants were intramedullarily implanted in the cut tibiae of cats to provide internal fixation (procedure shown in Fig. 6.6). An ultra-high-molecular-weight polyethylene (UHMWPE) rod, which does not exhibit

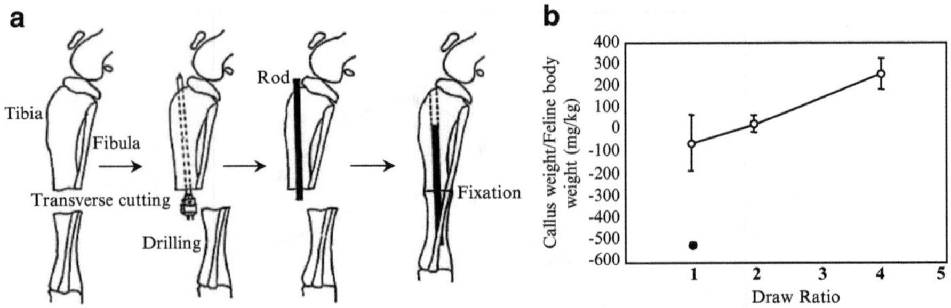

Fig. 6.6 (**a**) Procedure for implantation of PLLA rods into the intramedullary canal of sectioned cat tibiae. (**b**) The effect of the draw ratio of PLLA rods on the weight of callus formation at 8 weeks after implantation ($n=4$) PLLA(*open circle*), UHMWPE(*filled circle*) [10]

Fig. 6.7 Examples of PLLA screws, rods, and miniplates

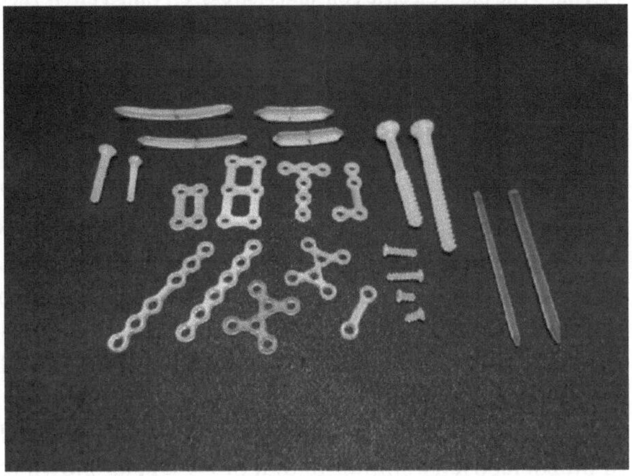

piezoelectricity, was used as a control. After 8 weeks, enhanced fracture healing was clearly observed for the PLLA rods with high draw ratios. These rods promoted larger callus formation than undrawn PLLA or UHMWPE rods.

Tajitsu et al. developed electrically controlled PLLA fibers by using the concept of piezoelectricity for applications in tools such as forceps [11]. PLLA fibers with high crystallinity and high orientation of crystallites were fabricated using high-speed spinning. When either DC or AC voltage was applied to a pair of PLLA fibers, they moved like forceps.

Screws, Pins, and Rods

Drawn PLLA and SR-PLLA were first used for bone fixation in the form of screws, pins, and rods. Different sizes and shapes of these materials are currently available in the market (Fig. 6.7).

Matsusue et al. investigated the degradation of drawn PLLA in vitro and in vivo [12]. Ultra-high-strength PLLA rods were fabricated using a drawing technique. Rods with a diameter of 3.2 mm and a draw ratio of 2.5:1 showed initial bending strength and modulus values of 240 MPa and 13 GPa, respectively. The weight of PLLA rods in the medullary cavity was reduced by 22% at 52 weeks and by 70% at 78 weeks after implantation in rabbits. Histologically, neither inflammatory nor foreign-body reaction was observed in the medullary cavity after 52 weeks. A bending strength greater than that of the human cortical bone was maintained for 8 weeks in the medullary canal.

Laine et al. summarized 10 years of clinical studies of SR-PLLA and SR-PDLLA screws and miniplates used in orthognathic surgery [13]. For 163 patients who had undergone a total of 329 orthognathic osteotomies fixed with PLA devices, the outcome was generally excellent, and very few minor complications (8.6%) were found, none of which affected the end result of the operation.

The mechanical strength of bioabsorbable materials is a concern in the stabilization of mandible fractures, where constant movement results in high force. The most common procedure for mandibular advancement (i.e., moving the lower jaw position forward) is bilateral sagittal split osteotomy (SSO). Maurer et al. tested four screws (made of PLLA or LLA copolymers) that are used clinically for fixation in SSO [14]. All four screws were found to be sufficiently stable at the osteotomy gap to withstand chewing forces.

Oba et al. compared the stability of the mandible following surgical orthodontic SSO and orthodontic multibracket treatments that used drawn PLLA (FIXSORB-MX; Takiron, Japan) and titanium screws [15]. They found no significant differences in the stability of bony segments for 23 patients treated with PLLA screws and 22 patients treated with titanium screws. However, a slight tendency for clockwise rotation of the distal mandible segment was observed in patients with PLLA screws. Therefore, they suggest that the fixation of bony segments with PLLA screws after SSO may be effective only when properly selected.

In Japan and China, PLLA pins have been successfully used to close the sternum cut for cardiovascular surgery and the ribs cut for thoracic surgery because high strength is not required for the pins in these surgeries.

The unnecessarily long strength retention and degradation time of PLLA may be adjusted by copolymerization with other monomers or blending with other polymers such as PDLLA. Biomet Microfixation, a US company, offers the LactoSorb® plating system for craniomaxillofacial fixation, first introduced in 1996. Figure 6.8, represented by the company, indicates that LactoSorb® is bioabsorbed in approximately 1 year or less. This graph shows that at the time of bone union, typically 8 weeks, LactoSorb® plating still retains 70% of its initial strength. According to the company, LactoSorb® consists of a copolymer of 82% LLA and 18% GA monomer. If LactoSorb® were composed of a polymer blend of 82% PLLA homopolymer and 18% PGA homopolymer, part of the PLLA component would remain unabsorbed in the body more than 1 year after insertion. Pietrzak and Kumar compared the in vitro hydrolysis of poly (LLA-co-GA) with a LLA/GA ratio of 82:18 with that of poly(LLA-co-GA) with a LLA/GA ratio of 85:15 [16]. The result of tensile modulus

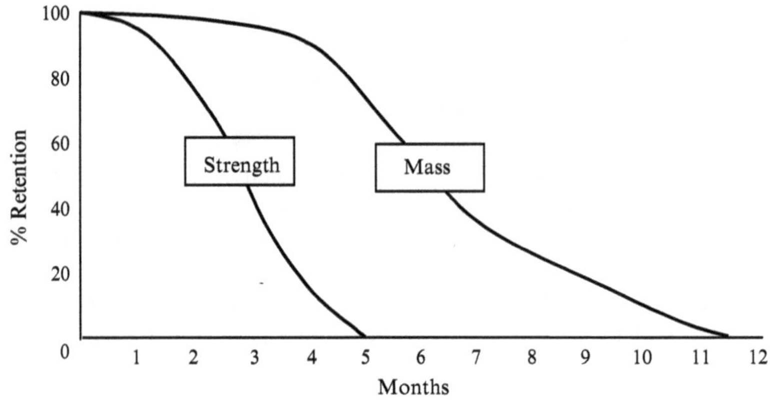

Fig. 6.8 Retention of mechanical strength and mass of LactoSorb® over time, expressed as a percentage of the original values

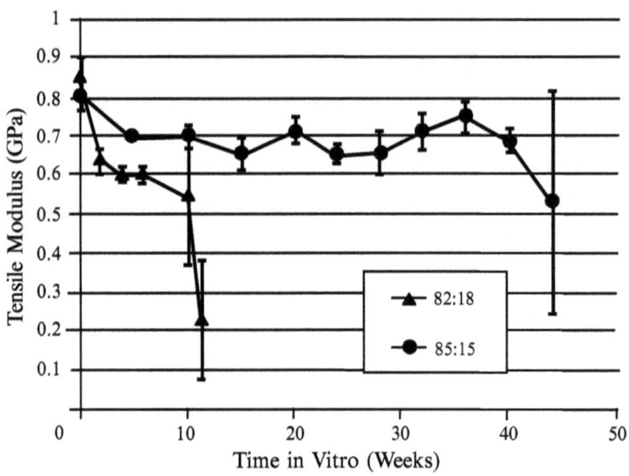

Fig. 6.9 Tensile modulus versus time at pH 7.4 and 37°C. Note that an errant (high) data point at 8 weeks for the 82:18 copolymer data set was removed for clarity. Average ± standard deviation [16]

is shown in Fig. 6.9. As can be seen, the copolymer (82:18) used for LactoSorb® lost the stiffness much faster than the copolymer (85:15), although the difference in chemical composition is not large. The remarkable difference in modulus change might be due to the different molecular weight of these copolymers. The inherent viscosity was 1.35 dL/g for the copolymer (82:18) and 1.88 dL/g for the copolymer (85:15). The mechanical properties and bioabsorption profile of the copolymer and the blend are very different, even if their chemical composition is similar. Biomet Microfixation likely uses unique proprietary processing techniques in plate manufacturing. LactoSorb® has been used successfully in more than 50,000 craniomaxillofacial cases since its introduction.

Despite these successes, there are some complications associated with these materials. Konan et al. reviewed adverse effects of bioabsorbable interference screws used in anterior cruciate ligament (ACL) fixation [17]. Although complications are rare, there have been reports of chondral damage caused by breakage of screws used in the ACL. Because screw heads remaining in the subcutaneous tissue may cause breakage, pain, or nonseptic foreign-body reactions, Sugimoto et al. suggested the simple removal of the PLLA screw head by using a micro-bone-saw as one way to prevent these problems [18].

Plates

The use of PLLA plates has been limited mainly to maxillofacial surgery, as they are not strong enough for load-bearing applications. Suzuki et al. studied a PLLA miniplate system for the treatment of mandibular condylar process fracture in 14 patients (aged 23.1 ± 5.7 years) [19]. Although two patients had mild chronic postoperative tenderness at the implantation site, no wound infection was observed. They found satisfactory bone healing in all patients, and there was no evidence of abnormal bioabsorption of the condylar process.

Ricalde et al. compared the strength of six different bioabsorbable implant plating systems before and after heating in an in vitro study [20]. A red oak wood board was cut into $12 \times 28 \times 160$ mm^3 blocks. The woodblocks were then cut in half. Six plates that differed in polymer composition and absorption time were chosen for this study on the basis of their popularity and similarity of use. The Inion system (Stryker Leibinger, Oklahoma City, OK) is composed of PLLA, PGA, PDLLA, and TMC, and its bioabsorption time is 1–3 years. The previously described LactoSorb® system (Lorenz, Jacksonville, FL) was the first in the market. Resorb X™ (KLS Martin, Jacksonville, FL) is composed of pure PDLLA. It is completely amorphous and retains its mechanical properties over a period of approximately 10 weeks before beginning the bioabsorption process, but complete bioabsorption time takes 1 year. Synthes (Synthes, West Chester, PA) is composed of 70% PLLA and 30% PDLLA. Its bioabsorption time is 2 years. Synthes Rapid (Synthes) is manufactured from 85:15 poly(LLA-co-GA) and has a bioabsorption time of 1 year. Not included in the study was the Bioabsorbable Fixation System (Osteomed, Addison, TX) because it has the same chemical composition as the Synthes product. The plates were cut to six holes and placed into either the heated or nonheated group (Table 6.2). After the plates were affixed to the oak blocks, a test machine automatically recorded the force and displacement behavior of each specimen ($n = 72$). The loads required to cause displacements of 1 mm and 2 mm are shown in Fig. 6.10. No statistically significant differences were found between loads required for 1- and 2-mm displacement in any of the specimens. The LactoSorb® plates performed significantly better than the other plates tested with regard to the force required to cause a clinically significant displacement of 1–2 mm. The LactoSorb® plates were uniformly superior at 2-mm displacement.

Table 6.2 Group contents[a] [20]

Group	Characteristics
Inion	Single 10-hole Inion (Stryker Leibinger) resorbable plate, cut to 6 holes, with six 2.0×7 mm resorbable screws per plate
LactoSorb®	Single 8-hole LactoSorb® (Lorenz) resorbable plate, cut to 6 holes, with six 2.0×7 mm resorbable screws per plate
Macropore	Single 8-hole Macropore (Medtronic) resorbable plate, cut to 6 holes, with six 2.0×6 mm resorbable screws per plate
Resorb X™	Single 8-hole Resorb X™ (KLS Martin) resorbable plate, cut to 6 holes, with six 2.1×7 mm resorbable screws per plate
Synthes	Single 8-hole Synthes (Synthes) resorbable plate, cut to 6 holes, with six 2.0×6 mm resorbable screws per plate
Synthes Rapid	Single 8-hole Rapid (Synthes) resorbable plate, cut to 6 holes, with six 2.0×6 mm resorbable screws per plate

[a]Specimens from each group were tested both as received from the manufacturer and after heating

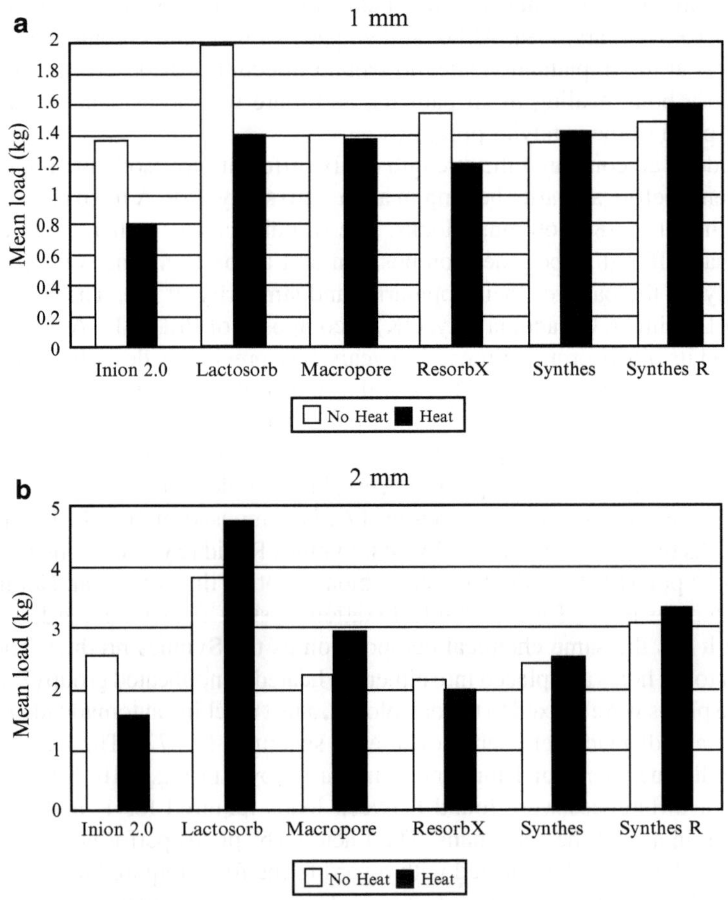

Fig. 6.10 Applied load required for 1-mm (**a**) and 2-mm (**b**) displacement [20]

PLLA implants also have applications in spine surgery [21]. Metallic spinal instrumentation and implants were developed to provide immediate structural stability until bone fusion occurs. However, metallic implants can cause several complications, including implant migration and failure, image degradation in MRI, and stress shielding.

A PLLA lumbar interbody cage device has been developed and effectively used for the treatment of patients with various degenerative disorders of the lumbar spine. PLLA cages packed with cancellous bone grafts have sufficient mechanical strength to support the spine directly after implantation. An in vivo study revealed that the rate of interbody fusion was higher 6 months after implantation of PLLA cages than after implantation of titanium cages. This increase in fusion is due to the decreased stiffness of the PLLA cages as compared to the titanium cages [22]. Under the high–load-bearing conditions of lumbar interbody fusion, the PLLA cage maintained its shape and height without collapsing, allowing fusion to occur [23].

When bioabsorption systems were first introduced in surgery, they tended to be made of materials selected more for their strength than their absorption rate. These implants were used in many clinical scenarios. With time, bioabsorbable products have become more popular for use in the pediatric population, and manufacturing companies have responded by creating "faster bioabsorbing" products. There is a delicate balance between strength and bioabsorption rates. Many clinicians and researchers have expressed concerns about trading strength for faster bioabsorption.

References

1. Cutright DE, Hunsuck EE, Beasley JD (1971) Fracture reduction using a biodegradable material, polylactic acid. J Oral Surg 29:393–397
2. Kulkarni RK, Pani KC, Neuman C, Leonard F (1966) Polylactic acid for surgical implants. Arch Surg 93:839–843
3. Yoshino N, Takai S, Watanabe Y, Kamata K, Hirasawa Y (1998) Delayed aseptic swelling after fixation of talar neck fracture with a biodegradable poly-L-lactide rod: case reports. Foot Ankle Int 19:634–637
4. Suganuma J, Alexander H (1993) Biological response of intramedullary bone to poly-L-lactic acid. J Appl Biomater 4:13–27
5. Eufinger H, Rasche C, Lehmbrock J, Wehmöller M, Weihe S, Schmitz I, Schiller C, Epple M (2007) Performance of functionally graded implants of polylactides and calcium phosphate/calcium carbonate in an ovine model for computer assisted craniectomy and cranioplasty. Biomaterials 28:475–485
6. Pietrzak WS (2000) Principles of development and use of bioabsorbable internal fixation. Tissue Eng 6:425–433
7. Matsusue Y, Yamamuro T, Yoshii S, Oka M, Ikada Y, Hyon S, Shikinami Y (1991) Biodegradable screw fixation of rabbit tibia proximal osteotomies. J Appl Biomater 2:1–12
8. Hanafusa S, Matsusue Y, Yasunaga T, Yamamuro T, Oka M, Shikinami Y, Ikada Y (1995) Biodegradable plate fixation of rabbit femoral shaft osteotomies. A comparative study. Clin Orthop Relat Res 315:262–271
9. Yasuda I, Noguchi K, Iida H (1955) Application to electrically-induced callus. J Jpn Orthop Assoc 29:351–353 (in Japanese)

10. Ikada Y, Shikinami Y, Hara Y, Tagawa M, Fukada E (1996) Enhancement of bone formation by drawn poly(L-lactide). J Biomed Mater Res 3:553–558
11. Tajitsu Y (2008) Piezoelectricity of chiral polymeric fiber and its application in biomedical engineering. IEEE Trans Ultrason Ferroelectr Freq Control 55:1000–1008
12. Matsusue Y, Yamamuro T, Oka M, Shikinami Y, Hyon SH, Ikada Y (1992) In vitro and in vivo studies on bioabsorbable ultra-high-strength poly(L-lactide) rods. J Biomed Mater Res 26: 1553–1567
13. Laine P, Kontio R, Lindqvist C, Suuronen R (2004) Are there any complications with bioabsorbable fixation devices? A 10 year review in orthognathic surgery. Int J Oral Maxillofac Surg 33:240–244
14. Maurer P, Holweg S, Knoll WD, Schubert J (2002) Study by finite element method of the mechanical stress of selected biodegradable osteosynthesis screws in sagittal ramus osteotomy. Br J Oral Maxillofac Surg 40:76–83
15. Oba Y, Yasue A, Kaneko K, Uchida R, Shioyasono A, Moriyama K (2008) Comparison of stability of mandibular segments following the sagittal split ramus osteotomy with poly-L-lactic acid (PLLA) screws and titanium screws fixation. Orthodontic Waves 67:1–8
16. Pietrzak WS, Kumar M (2009) An enhanced strength retention poly(glycolic acid)-poly(L-lactic acid) copolymer for internal fixation: in vitro characterization of hydrolysis. J Craniofac Surg 20:1533–1537
17. Konan S, Haddad FS (2009) A clinical review of bioabsorbable interference screws and their adverse effects in anterior cruciate ligament reconstruction surgery. Knee 16:6–13
18. Sugimoto K, Takakura Y, Tanaka Y, Kawate K (2003) Technique tip: fixation of Mitchell's osteotomy using a PLLA screw. Foot Ankle Int 24:372–373
19. Suzuki T, Kawamura H, Kasahara T, Nagasaka H (2004) Resorbable poly-L-lactide plates and screws for the treatment of mandibular condylar process fractures: a clinical and radiologic follow-up study. J Oral Maxillofac Surg 62:919–924
20. Ricalde P, Caccamese J, Norby C, Posnick JC, Hartman MJ, von Fraunhofer JA (2008) Strength analysis of 6 resorbable implant systems: does heating affect the stress–strain curve? J Oral Maxillofac Surg 66:2493–2497
21. Vaccaro AR, Singh K, Haid R, Kitchel S, Wuisman P, Taylor W, Branch C, Garfin S (2003) The use of bioabsorbable implants in the spine. Spine J 3:227–237
22. van Dijk M, Smit TH, Sugihara S, Burger EH, Wuisman PI (2002) The effect of cage stiffness on the rate of lumbar interbody fusion: an in vivo model using poly(L-lactic Acid) and titanium cages. Spine 27:682–688
23. van Dijk M, Tunc DC, Smit TH, Higham P, Burger EH, Wuisman PI (2002) In vitro and in vivo degradation of bioabsorbable PLLA spinal fusion cages. J Biomed Mater Res 63:752–759

Chapter 7
Growth Factors for Promoting Wound Healing

Abstract The wound-healing process requires many types of growth factors, and exogenous delivery of appropriate growth factors to the injured site leads to accelerated healing. To deliver such growth factors, carriers which are generally made of bioabsorbable polymers are employed. However, only a few release systems, including a combination of bone morphogenetic protein (BMP) and a collagen sponge, are commercially available for the delivery of growth factors. However, this collagen carrier needs a large amount of BMP due to the low efficacy of the carrier. Studies on more effective carriers for the growth factors are ongoing. Recently, technologies to concentrate the patient's own platelets which contain a large amount of the growth factors have been developed. Since the growth factors are completely autologous, it should be safe and cost effective. This chapter offers a review of recent studies on growth factor–carrier systems, as well as emerging technologies for platelet-rich plasma (PRP). Importance of carriers is emphasized to effectively deliver growth factors.

Introduction

As described in Chap. 2, the natural process of wound healing is triggered upon tissue injury. Wound healing is a complex and dynamic process that results in the restoration of anatomical continuity and function. Numerous cytokines and growth factors are involved in regulating the wound-healing process. They include epidermal growth factor (EGF), transforming growth factor-α,β (TGF-α,β), hepatocyte growth factor (HGF), vascular endothelial growth factor (VEGF), platelet-derived growth factor (PDGF), basic fibroblast growth factor (bFGF), and keratinocyte growth factor (KGF). Endogenous growth factors are produced by cells at the wound site. However, if their concentration is lower than required, a sufficient amount of growth factors should be exogenously supplied for tissue reformation. In this chapter, clinically available applications for delivering growth factors for wound healing are discussed, including drug-delivery systems (DDSs) for recombinant human growth factors and delivery of multiple growth factors using platelet-rich plasma (PRP) technology.

S. Suzuki and Y. Ikada, *Biomaterials for Surgical Operation*,
DOI 10.1007/978-1-61779-570-1_7,

DDS for Growth Factors

Many recombinant growth factors and cytokines are available for research purposes, but only a few growth factors have been shown to be effective enough in clinical studies to be approved for use in human therapies. Although some studies have concluded that growth factors are useful for tissue regeneration, others have not supported the application. This discrepancy may be due to the different administration routes of growth factors. If a growth factor is administered at the site of tissue formation in solution form (bolus administration), it diffuses away rapidly from the site of injection, resulting in low efficacy of the growth factor. Therefore, DDS strategies that control the release of the protein drugs locally for an optimal period of time will be required to achieve the desired outcome.

Growth Factors

Growth factors are signaling polypeptides that elicit specific cellular responses in a biological environment. Specific growth factors trigger specific cellular responses, including migration, differentiation, proliferation, cell survival, and apoptosis. Popular growth factors used in tissue formation research are listed in Table 7.1.

Bone morphogenetic proteins (BMPs) represent a family of related osteoinductive polypeptides akin to differentiation factors. In 1965, Urist discovered that

Table 7.1 Growth factors and their effects

Growth factor	Abbreviation	Effects
Basic fibroblast growth factor	bFGF, FGF-2	Angiogenesis; fibroblast and osteoblast mitogen
Bone morphogenetic proteins	BMP-2 BMP-7 (OP-1)	Growth and development of some tissues; osteogenesis
Transforming growth factor-β1	TGF-β1	Proliferation and differentiation of bone forming cells; fibroblast matrix synthesis
Vascular endothelial growth factor	VEGF	Angiogenesis; proliferation and migration of endothelial cells
Platelet-derived growth factor	PDGF	Proliferation of smooth muscle cells; fibroblast mitogen and matrix synthesis
Hepatocyte growth factor	HGF	Hepatocyte mitogen; mitogen and antiapoptotic factor of cells; angiogenesis
Keratinocyte growth factor	KGF	Epithelization of wounds
Epidermal growth factor	EGF	Proliferation of epithelial cells; mesenchymal and fibroblast cells
Insulin-like growth factor	IGF-1	Cartilage development and homeostasis; bone formation

demineralized bovine bone matrix was capable of inducing endochondral bone formation when implanted ectopically into soft tissues of experimental animals [1]. Urist and his colleagues subsequently discovered that glycoproteins with low molecular weight (MW) that were isolated from bone were responsible for the osteogenic activity observed earlier and were capable of inducing bone formation when delivered to ectopic or orthotopic locations. Wozney et al. and Celeste et al. subsequently cloned the first bone morphogenetic proteins: BMP-2, BMP-3, and BMP-4. Ozkaynak et al., using similar techniques, cloned BMP-7 and BMP-8. Currently, at least 20 BMPs have been identified, many of which can induce chondroosteogenesis in various mammalian tissues. BMPs are homodimeric molecules (MW 25–30 kDa) that regulate various cellular functions such as bone induction, morphogenesis, chemotaxis, mitosis, hematopoiesis, cell survival, and apoptosis. BMP-2 and BMP-7 (also known as osteogenic protein-1; OP-1) have been developed for clinical use and tested in a number of orthopedic, dental, and maxillofacial applications [2, 3]. Among the plethora of bioactive factors available, BMP-2 may be the strongest osteoinductive factor administered therapeutically to restore the form and function of bone.

Fibroblast growth factors (FGFs) induce fibroblast proliferation. Basic FGF (bFGF, also known as FGF-2) induces proliferation of endothelial cells, chondrocytes, smooth muscle cells, melanocytes, and others. It can also promote adipocyte differentiation, induce macrophage and fibroblast IL-6 production, stimulate astrocyte migration, and prolong neural survival. FGFs are potent modulators of cell proliferation, motility, differentiation, and survival, and play an important role in normal regeneration processes in vivo, such as embryonic development, angiogenesis, osteogenesis, chondrogenesis, and wound repair. Heparin enhances the mitogenic activity of bFGF and serves as a cofactor promoting the binding of bFGF to high-affinity receptors.

Vascular endothelial growth factor (VEGF) is the only known cytokine with mitogenic effects primarily confined to endothelial cells. VEGF is produced by a variety of normal and tumor cells. Its expression correlates with periods of capillary growth during embryonic development, wound healing, and the female reproductive cycle, as well as with tumor expansion. VEGF is thought to be a major promoter of both physiological and pathological angiogenesis. Furthermore, VEGF has been shown to be an antiapoptosis survival factor for endothelial cells even during periods of microvessel stasis. In addition, VEGF can enhance tissue secretion of a variety of proangiogenic proteases, including urokinase-type plasminogen activator (uPA), matrix metalloproteinase-1 (MMP-1), and matrix metalloproteinase-2 (MMP-2). It has also been shown that inhibition of VEGF or the VEGF-R2 receptor can suppress expression of MMP-2 and MMP-3.

TGF-β1 belongs to a superfamily of growth and differentiating factors of which BMP is a member. TGF-β1 (MW 25 kDa) is synthesized in platelets and macrophages, as well as in some other cell types. When released upon platelet degranulation or actively secreted by macrophages, TGF-β1 acts as a paracrine growth factor (i.e., growth factor that is secreted by one cell and exerts its effect on an adjacent cell), mainly affecting fibroblasts, marrow stem cells, and osteoblast precursors.

TGF-β1 stimulates the chemotaxis and mitogenesis of osteoblast precursors (resulting in a local increase in populations of healing cells), promotes differentiation toward mature osteoblasts, stimulates deposition of collagen matrix, and inhibits osteoclast formation and bone resorption. TGF-β1, therefore, represents a long-term, sustained mechanism for healing and bone formation.

Platelet-derived growth factor (PDGF) is a glycoprotein existing mostly as a dimer of two chains of about equal size and MW (14–17 kDa). This protein appears to be the first growth factor present in a wound and is synthesized not only in platelets but also in macrophages and endothelium. It initiates connective tissue healing, including bone regeneration and repair. At the time of injury, PDGF emerges from degranulating platelets and activates cell membrane receptors on target cells. The most important specific activities of PDGF include mitogenesis, angiogenesis (endothelial mitosis into functioning capillaries), and macrophage activation (debridement of the wound site and a second-phase source of growth factors for continued repair and bone formation). PDGF is also stored in the bone matrix and released upon activation of osteoblasts, resulting in an increase in new bone formation. PDGF is a potent stimulator of fibroblast cell migration, mitogenesis, proliferation, and matrix synthesis, all of which are important in wound healing.

Delivery of Growth Factors

The use of growth factors has not always been successful in vivo. The major reasons for this are the high diffusibility and short half-life of growth factors in vivo. Hence, effective delivery systems are required to locally sustain growth factor concentrations and to enhance their activities. Delivery methods include bolus injection; release of protein adsorbed directly on the carrier surface, in collagen sponge, or in porous coating; constant delivery via osmotic pump; and controlled release of protein trapped in a carrier made of absorbable polymers. It is very important for the growth factor carrier to be able to capture growth factors without impairing their biological activity and release them at the desired time schedule [4].

As illustrated in Fig. 7.1, with the use of an effective DDS, growth factors can be delivered to a specific site in a controlled manner with respect to both time period and release rate, without them reaching either toxic or ineffective dose levels [5, 6]. Numerous studies have been conducted using different combinations of growth factors and bioabsorbable polymers as drug-delivery vehicles or carriers, as listed in Table 7.2. Synthetic and natural polymers frequently used as carriers are poly (glycolic acid) and poly(lactic acid) and their copolymers, poly(ethylene glycol), alginate, gelatin, and collagen. Carriers take physical forms such as porous sponges, microspheres, micro- and nanocapsules, and hydrogels [7]. The release profile of growth factors can be modulated in two ways: by manipulating the polymeric carrier properties [8] or by adjusting the physical and chemical properties of the carrier, such as degradation rate, degree of cross-linking, and porosity [7]. Care should be

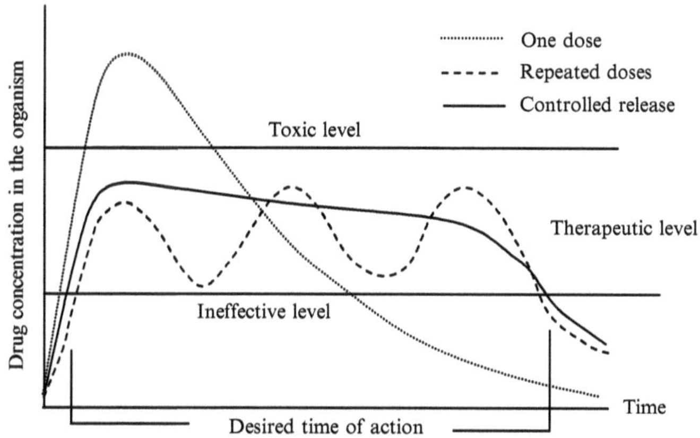

Fig. 7.1 Scheme of drug concentrations in the body over time when using different administration methods [5]

Table 7.2 Polymeric growth factor carriers

Growth factor	Carrier	Animal	Tissue regenerated
rhBMP-2	PLA	Dog	Long bone
	PLA microsphere	Rabbit	Skull bone
	PGLA microsphere	Rat	Skull bone
	Gelatin	Rabbit	Skull bone
		Dog	Cartilage, Skull bone
	Gelatin-calcium phosphate cement	Goat	Long bone
	Alginate-PCL mesh	Rat	Long bone
EGF	PVA	Rat	Dermis
	Gelatin	Rat	Dermis
bFGF	Alginate	Mouse	Angiogenesis
	Gelatin	Mouse	Angiogenesis, Dermis
		Rabbit, Monkey	Skull bone
		Dog	Nerve
	Collagen	Mouse	Cartilage
	Chitosan	Mouse	Dermis
	Flagmin-protamine microsphere	Mouse	Angiogenesis
PDGF-BB	Fibrin-heparin	Dog	Tendon
IGF-1	Fibrin gel	Horse	Cartilage
TGF-β1	Alginate	Rabbit	Cartilage
	PEG	Rat	Dermis
VEGF	Alginate	Mouse	Angiogenesis
	Collagen	Rat	Heart
IGF-1/bFGF	PGLA-PEG	Rat	Adipogenesis
VEGF/PDGF	PGLA scaffold	Mouse	Angiogenesis
VEGF/IGF-1	Alginate	Mouse	Angiogenesis, Muscles

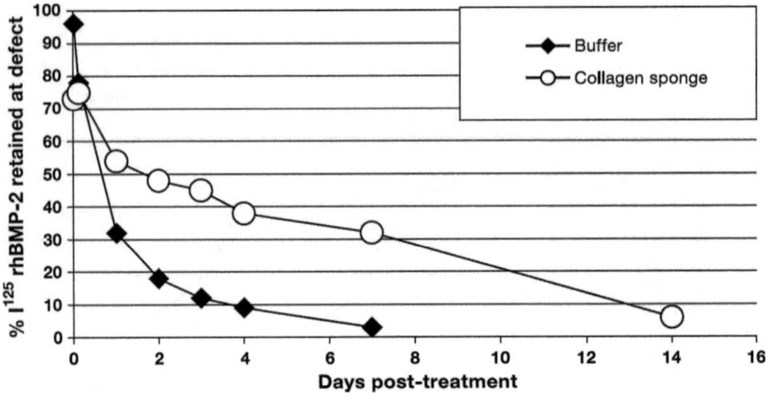

Fig. 7.2 Retention of rhBMP-2 in a rabbit ulna osteotomy model, implanted with and without a collagen sponge [9]

taken when organic solvents and cross-linkers are employed in the formulation of the growth factor/carrier mixture because of possible denaturation of proteins by these reagents, which would negatively affect protein association and function.

Commercially Available Growth Factor–Delivery Systems (BMP-Collagen)

The BMPs that have been approved for use in humans utilize collagen for their delivery. BMP-2 with collagen sponge as a carrier (INFUSE® Bone Graft Medtronic Sofamor Danek, Memphis, TN, USA; InductOs™, Wyeth Europa, Maidenhead Berkshire, UK) was approved for interbody fusion procedures in 2002, for open tibial fractures in 2004, and for alveolar ridge and sinus augmentations in 2007 [3]. BMP-7 in combination with collagen (Osigraft™, Howmedica International S.de R.L. Raheen, Limerick, Ireland) was approved for long bone fractures and as an alternative to autografts in patients requiring posterolateral lumbar spinal fusion [2].

Figure 7.2 shows the release profile of radiolabeled recombinant human BMP-2 (^{125}I-rhBMP-2) soaked in a collagen sponge and implanted in a rabbit ulna osteotomy model [9]. The local retention was prolonged to a significantly greater extent than the retention in the solution delivery. Seven days after implantation, about 37% of the initial dose remained at the site, whereas only 3% remained for bolus injection.

Both INFUSE® Bone Graft and InductOs™ systems consist of a solution of rhBMP-2 and an absorbent collagen sponge (ACS). The amount of BMP-2 in INFUSE® Bone Graft is 4.9 or 12.7 mg (1~1.3 mg/mL), while that of InductOs™ is 12 mg (1.5 mg/mL). These systems are used in anterior lumbar interbody fusion (ALIF) procedures in combination with a titanium cage implant (called LT-CAGE® Lumbar Tapered Fusion Device). For preparation, rhBMP-2 powder is dissolved in sterile water, and a collagen sponge is soaked in the rhBMP-2 solution for at least 15 min. The sponge is then rolled up and placed inside a titanium cage. The damaged disc from the patient's spine is removed, and the cages containing the collagen sponges soaked with rhBMP-2 are implanted. The device stabilizes the spine while fusion occurs, and maintains proper spacing between the vertebrae. It is an alternative to using autograft bone and eliminates the pain and complications associated with harvesting the iliac crest. A clinical trial of 679 patients undergoing anterior lumbar fusion was conducted [10]. Two hundred seventy-seven of these patients received implantation of the INFUSE® Bone Graft/LT-CAGE® Lumbar Tapered Fusion Device, while 402 underwent an autograft transfer from the iliac crest. The patients treated with BMP-2 had statistically superior outcomes with regard to the length of surgery, blood loss, hospital stay, reoperation rate, median time to return to work, and fusion rates at 6, 12, and 24 months. A randomized, controlled trial, with 15 patients in each group, was carried out to compare the effect of rhBMP-2/collagen sponge in combination with freeze-dried cancellous allograft, with autogenous bone graft, for reconstruction of diaphyseal tibial fractures with cortical defects [11]. The clinical results at the 1-year postoperative follow-up for rhBMP-2/allograft-treated patients were comparable with the results of autogenous bone grafting.

OP-1 Implant™ is a commercial product of rhBMP-7 in combination with bovine collagen powder. Each vial of OP-1 Implant™ (Olympus Biotech, Hopkinton, MA) contains 3.3 mg of rhBMP-7 and 1 g of bovine collagen powder, while OP-1 Putty™ also contains CMC (230 mg) as an additive, which makes it malleable and easy to handle [12]. OP-1 Putty™ is approved as an alternative to grafting in patients requiring nonfeasible bone marrow harvest and revision to posterolateral lumbar spinal fusion. Prior to use, Osigraft is reconstituted with 2–3 mL of saline (0.9% w/v) solution to form a slurry. It is used as bone graft material for long-bone nonunion.

The efficacy of OP-1 Implant™ was evaluated in a multicenter clinical trial for the treatment of tibial nonunions of 122 patients [13]. Patients were treated with insertion of an intramedullary rod, accompanied by OP-1 Implant™ (63 patients) or a fresh bone autograft (61 patients). Nine months after the insertion, 81% of the OP-1-treated patients were clinically successful, compared to 85% of patients who received autografts. After 2 years, the clinical success rates in the 2 groups continued to be similar, with no statistically significant difference ($p=0.939$). It was concluded that OP-1 Implant™ was a safe and effective treatment for tibial nonunions.

Another clinical trial was conducted to evaluate the safety and efficacy of OP-1 Putty™ for single-level uninstrumented posterolateral fusion of the lumbar spine following decompressive laminectomy for the treatment of symptomatic degenerative spondylolisthesis with spinal stenosis [14]. Thirty-six patients were randomized to either the OP-1 Putty™ (24 patients) or the autogenous iliac crest bone graft (12 patients) groups. Clinical and radiographic outcomes evaluated after 1 year revealed the success rate of OP-1-fused patients to be 86%, compared to 73% for autograft-fused patients. OP-1 Putty™ was found to be safe since there were no adverse consequences related to this implant in the patient population. After 2 and 3 years postoperatively, 55% of OP-1 patients showed radiographically solid fusion, compared with a 40% fusion rate in autograft patients [15].

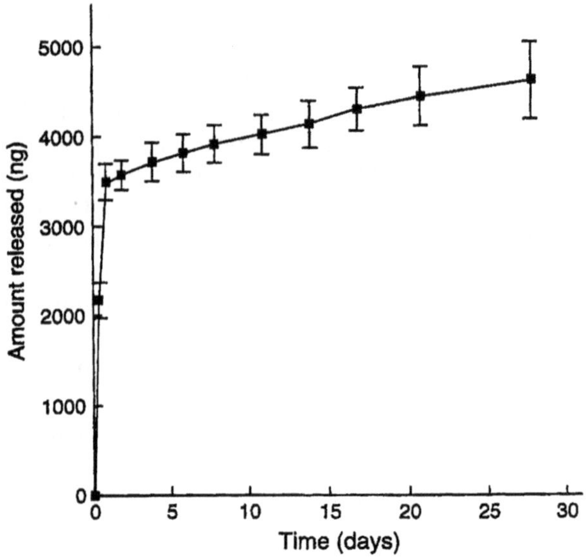

Fig. 7.3 Cumulative release of rhBMP-2 from PLLA/TCP membranes in vitro [16]

Notwithstanding the remarkable clinical success for the commercially available BMP-collagen carrier system, the fact remains that the concentration of BMP used is a million times higher (milligrams) than in vivo concentrations (nanograms). Therefore, materials such as fibrin glue and ceramics have been clinically used as alternative BMP-2 carriers, but they are less than ideal because of the poor capability of these biomaterials to control the release of BMP. It is also important to capture the growth factors without deactivating them. The search for effective delivery systems for biologically active agents is therefore still ongoing.

Developments in Drug-Delivery Systems

The typical pattern of growth factor release from carriers consists of two phases: an initial burst phase followed by a slow-release phase. Lee et al. investigated the release profile of rhBMP-2 from stiff PLLA/tricalcium phosphate (TCP) membranes [16]. The membranes were soaked in rhBMP-2 solution to allow loading of 5 μg of BMP. Each membrane was then placed in PBS solution at 37°C, and aliquots of the solution were collected at predetermined times. The concentration of BMP-2 was measured using an enzyme-linked immunosorbent assay (ELISA), and the result of cumulative release kinetics is shown in Fig. 7.3. During the first day (initial burst phase), approximately 70% of BMP was released (3.5 μg after 24 h). After that, the rhBMP-2 was consistently released at a rate of 7–10 ng/day for up to 4 weeks (slow-release phase). They also studied the effect of this rhBMP-2-loaded membrane on bone augmentation in a rabbit calvarial model and found that it enhanced bone augmentation to a significantly greater extent than a control membrane

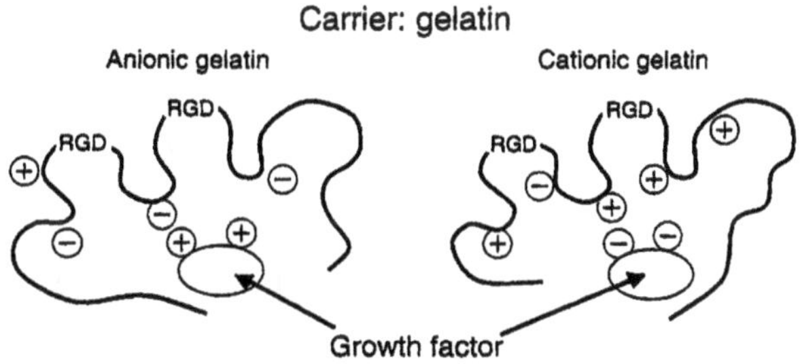

Fig. 7.4 Entrapping growth factors using biodegradable gelatin [17]

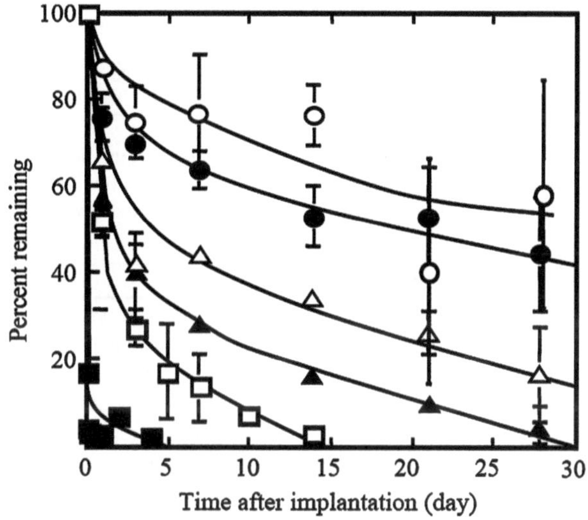

Fig. 7.5 In vivo time profiles of residual radioactivity following subcutaneous implantation of gelatin hydrogels incorporating ^{125}I-labeled BMP-2 into the backs of mice. The hydrogel water contents are (*open circle*) 93.8%, (*filled circle*) 96.9%, (*open triangle*) 97.8%, (*filled triangle*) 99.1%, and (*white square*) 99.7%. The symbol (*black square*) indicates residual radioactivity after injection of ^{125}I-labeled BMP-2 [18]

lacking rhBMP-2. Since the rhBMP-2 released initially quickly diffuses out from the site, such a high dose of rhBMP-2 might be necessary for bone regeneration in this model.

Growth factors can be incorporated into oppositely charged carriers by electrostatic interaction without degradation. Gelatin molecules interact with growth factors through ionic interaction, as shown in Fig. 7.4 [17]. They are released from the gelatin carrier as a result of enzymatic degradation of gelatin. The equilibrium water content of gelatin hydrogels markedly affects the release kinetics. To investigate the release profile of BMP-2 from gelatin carriers, gelatin hydrogels incorporating ^{125}I-labeled BMP-2 were subcutaneously implanted into the backs of mice, and the residual radioactivity of the hydrogels and surrounding tissues was measured using a gamma counter at predetermined times after implantation [18]. Figure 7.5 shows the release profiles of BMP-2 from gelatin hydrogels with water contents ranging from 93.8 to 99.7%. The effect of BMP-2 loaded in gelatin hydrogels on ectopic bone formation was evaluated. Gelatin hydrogels incorporating 0, 0.5, 1, and 5 μg of BMP-2 were implanted subcutaneously. The skin tissues around the implantation site were removed 1, 2, and 4 weeks later, and the alkaline phosphatase (ALP) activity of the tissues was measured using the *p*-nitrophenyl-phosphate method. It was

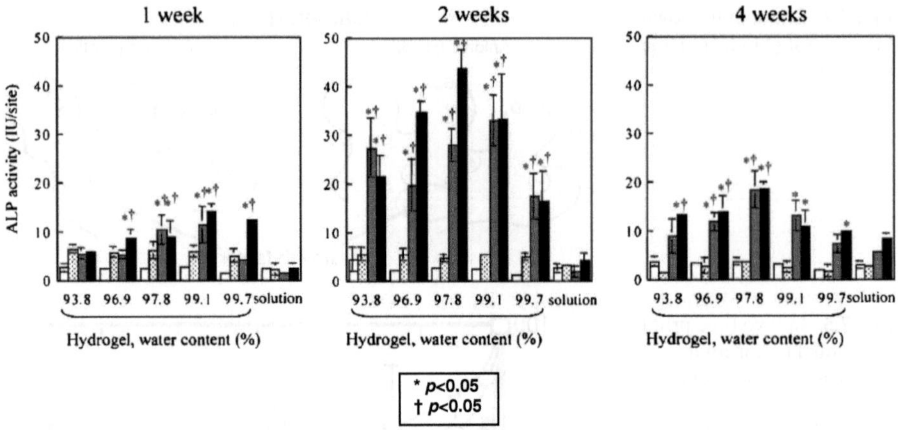

Fig. 7.6 ALP activity of tissues around the implantation site of gelatin hydrogels incorporating BMP-2, or the injection site of BMP-2 solution, after 1, 2, and 4 weeks. The BMP-2 doses are (*white square*) 0, (*light grey square*) 0.5, (*dark grey square*) 1, and (*black square*) 5 μg [18]

Table 7.3 Stress required for breakage of anterior tracheal cartilage [19]

	Stress (10^6 N/m^2)
Control group	0.68±0.43*†
Gelatin sponge group	1.68±1.26‡§
BMP group	4.92±1.32**
Normal trachea	3.52±1.03

Data are represented as mean±SD

*$p=0.0001$ vs. gelatin sponge group
†$p<0.001$ vs. normal trachea
‡$p=0.0015$ vs. BMP group
§$p=0.002$ vs. normal trachea
**$p=0.013$ vs. normal trachea

found that gelatin hydrogels incorporating 1 and 5 μg of BMP-2 enhanced the ALP activity of subcutaneous tissues to a significantly greater extent than the treatment with BMP-2 solution alone. The gelatin hydrogel containing 5 μg of BMP-2 with a water content of 97.8% showed the highest ALP activity 4 weeks after implantation, as represented in Fig. 7.6.

Okamoto et al. investigated whether BMP-2, released slowly from a gelatin sponge, could induce cartilage regeneration in a canine model of tracheomalacia and evaluated the long-term results [19]. A gap was created in the anterior cervical trachea by removing ten sequential strips of cartilage, each 5 cm long. In the control group, the gaps were left untreated, while in the gelatin sponge group, a gelatin sponge soaked in buffer solution was implanted in each defect. In the BMP group, a gelatin sponge soaked in buffer solution containing 12 μg of BMP-2 was implanted in each defect. No regenerated cartilage was detected in the control or gelatin sponge group, even 6 months after surgery. In contrast, cartilage was regenerated from the host perichondrium, around the stumps of the resected cartilages, in the BMP group. This regenerated cartilage maintained the integrity of the internal lumen for longer than 6 months. The result of a compressive fracture test (Table 7.3) revealed that the

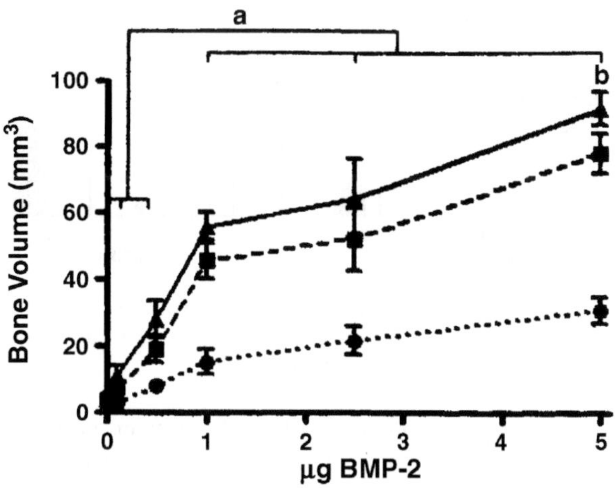

Fig. 7.7 Micro-CT evaluation of bone volume as a function of rhBMP-2 dose at week 4 (*light dashed lines*), week 8 (*bold dashed lines*), and week 12 (*solid lines*). (**a**) $p<0.05$ as indicated, (**b**) $p<0.05$ versus all other groups [21]

tracheal cartilage in the BMP group was significantly more stable than that in the gelatin sponge and control groups.

The gelatin hydrogel, a carrier for slow release of BMP-2, was coupled with a lactide-based biodegradable copolymer and grafted into bone defect sites of a canine orbital floor fracture model, a model for bone healing of difficult defects [20]. The release profile of radio-iodinated BMP-2 from the gelatin hydrogel carrier subcutaneously implanted into nude mice was similar to that of the radio-labeled hydrogel carrier alone (~60% at day 3 and ~80% at day 14). A bolus injection of BMP-2 resulted in 90% loss at the injection site within the first 8 h. In the canine orbital floor fracture model, new bone formation and defect healing at 5 weeks were enhanced to a greater extent with the BMP-2 hydrogel carrier system than with the use of copolymer saturated with the same amount of BMP-2 without gelatin hydrogel.

Boerckel et al. investigated the effects of rhBMP-2 dose on the regeneration of critically sized rat femoral bone defects over 12 weeks for the hybrid nanofiber mesh (PCL)/alginate delivery system [21]. For loading rhBMP-2 into the delivery system, the rhBMP-2 solution was mixed at six different concentrations with arginine-glycine-aspartic acid (RGD)-functionalized alginate to a final concentration of 2% alginate, then the mixture was cross-linked rapidly with 0.84% $CaSO_4$. The bilateral, critically sized (8 mm) segmental defects were surgically created in femora of 13-week-old Sprague Dawley rats. For the hybrid system group, a nanofiber mesh tube was fitted over the bone ends, and RGD-functionalized alginate hydrogel (200 μl) containing 0.0, 0.1, 0.5, 1.0, 2.5, and 5.0 μg of rhBMP-2 was injected into the defect space. For the collagen sponge group, rhBMP-2 was pipetted onto the sponge (8×5 mm^2) 10 min prior to implantation at either 2 or 20 μg/ml for the 0.1- and 1.0-μg groups, respectively, and they were then press fitted into the critically sized defect. Micro-CT was used to qualitatively demonstrate the bone formation. For the hybrid delivery system, the bone formation responded in a nonlinear, dose-dependent manner to rhBMP-2 (Fig. 7.7). At week 12, the 1.0 and 2.5 μg doses had significantly greater bone volume than those equal or under 0.5 μg doses, and 5 μg

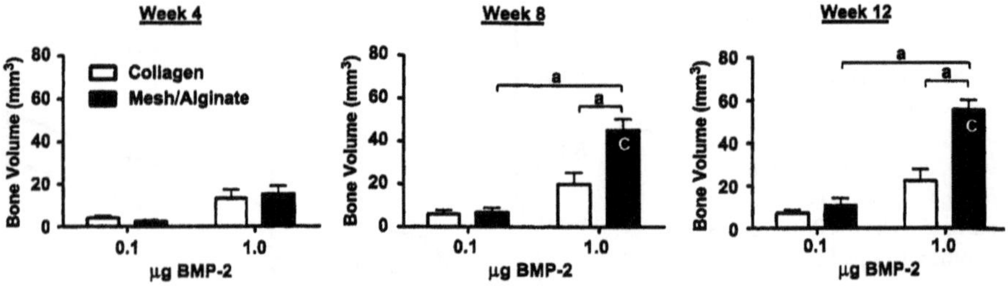

Fig. 7.8 Micro-CT quantification of bone volume in the nanofiber mesh/alginate (*filled square*) and collagen groups (*open square*) at weeks 4, 8, and 12 postsurgery. *a* $p<0.05$ as indicated, *C* $p<0.05$ vs. same group at week 4 [21]

dose group exhibited significantly greater bone volume than all other groups. Figure 7.8 represents comparison of the hybrid and collagen sponge delivery systems with 0.1 or 1.0 μg rhBMP-2 on bone formation at 4, 8, and 12 weeks postoperatively. At weeks 8 and 12, the bone volume in the hybrid system with 1.0 μg of rhBMP-2 was significantly larger than that of the collagen sponge system containing the same amount of rhBMP-2.

Hankemeier et al. analyzed the effect of low-dose (3 ng/mL) and high-dose (30 ng/mL) bFGF treatment on the proliferation, differentiation, and apoptosis of human bone marrow stem cells (BMSCs), and compared the results with those of a control group without bFGF [22]. Low-dose bFGF triggered a biphasic BMSC response; on day 7, cell proliferation reached its maximum and was significantly higher than that of other groups. On day 14 or 28, expression of types I and III collagen, fibronectin, and α-smooth muscle actin mRNA was significantly enhanced in the presence of low-dose bFGF. In contrast, high-dose bFGF did not stimulate BMSC differentiation or proliferation. Vimentin mRNA was expressed in cultures treated with low-dose and high-dose bFGF after 14 and 28 days, respectively. Cell density, measured on days 7, 14, and 28 posttreatment, was significantly higher in cultures treated with low-dose bFGF than in those treated with high-dose bFGF. The apoptosis rate remained stable, at a rather high level, in all groups. Microscopic investigation of the cell cultures treated with low-dose bFGF showed homogeneous, dense, fibroblast-like, spindle-shaped cells with long cell processes, unlike cultures treated with high-dose or no bFGF.

The ability of water-insoluble fragmin/protamine microparticles (diameter of 0.5–1 μm) to control the release of bFGF was evaluated in vitro and in vivo [23]. Fragmin is a low-MW heparin, which has much lower anticoagulant activity than native heparin. It also has the ability to bind heparin-binding growth factors such as FGFs, HGFs, VEGFs, TGFs, and PDGFs. Protamine is a purified mixture of proteins obtained from fish sperm. It neutralizes heparin to form a complex lacking anticoagulant activity. The microparticles were synthesized by adding protamine solution (10 mg/mL) to fragmin solution (6.4 mg/mL) under hard vortexing, and the resultant microparticles were washed with PBS. To investigate the release profile of bFGF, ^{125}I-bFGF was added to the microparticles in the presence of BSA for 30 min at room temperature, and the sample was incubated in culture medium.

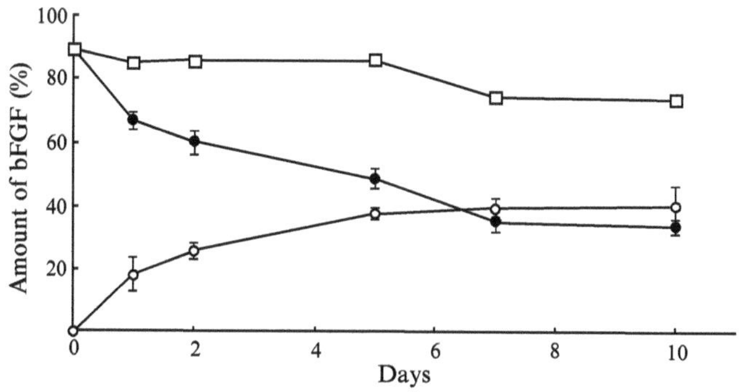

Fig. 7.9 The release profile of bFGF from fragmin/protamine microparticles. The amounts of bFGF released into DMEM at 37°C (*open circle*) and retained within the microparticles (*filled circle*), and the sum of those 2 amounts (*open square*) [23]

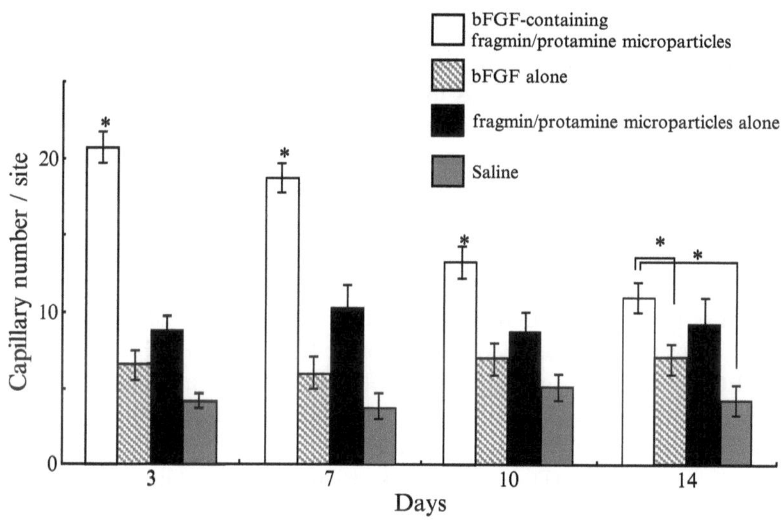

Fig. 7.10 Effects of injection of bFGF containing fragmin/protamine microparticles on vascularization in vivo [23]

As shown in Fig. 7.9, about 15% of ^{125}I-bFGF was released from the particles within 1 day, followed by gradual release. Microparticles incorporating bFGF were subcutaneously injected into mice, and vascularization was evaluated from the number of capillaries observed from microphotographs of each section ($n=6$). The number of capillaries observed per site from the histological examination is shown in Fig. 7.10. Fragmin/protamine microparticles were found to be an effective carrier for bFGF.

Therapeutic vascularization remains a significant challenge in tissue formation. Phelps et al. synthesized PEG-based hydrogel matrices presenting protease-degradable

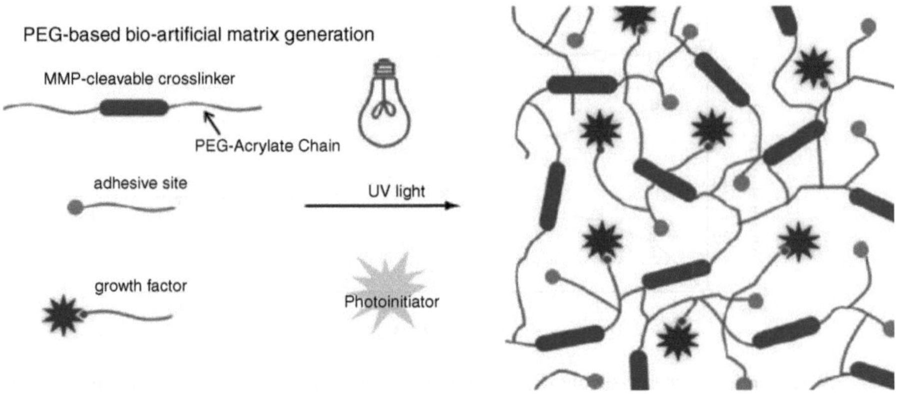

Fig. 7.11 Design of PEGDA-based bioartificial matrices [24]

sites, cell-adhesion motifs, and a growth factor (i.e., VEGF), to induce the growth of vasculature in vivo [24]. Figure 7.11 shows the study design: Matrix metalloproteinase (MMP)-degradable PEGDA macromer (A-PEG-GPQ-PEG-A) was synthesized by reacting the proteolytically cleavable peptide GPQGIWGQK with acrylate-PEG_{3400}-NHS, a primary amine reactive cross-linker. PEG-acrylate functionalized with arginine-glycine-aspartic acid tripeptide (RGD) adhesive ligand (A-PEG-RGD) was synthesized by reacting acrylate-PEG_{3400}-NHS with the cell adhesive peptide GRGDSPC. PEG-acrylate functionalized with VEGF (A-PEG-VEGF) was synthesized by reacting acrylate-PEG_{3400}-maleimide, a sulfo-hydryl-reacting cross-linker, with VEGF121-cys (VEGF containing an additional C-terminal cysteine). Hydrogel matrices consisting of A-PEG-GPQ-PEG-A, A-PEG-RGD, and A-PEG-VEGF were generated by polymerizing the acrylate end-functional groups with low-intensity UV light in the presence of a photoinitiator.

For implantation, hydrogels were photopolymerized in cylindrical silicone molds around polycaprolactone (PCL) mesh disks to allow for visualization and retrieval of implants. Hydrogels were implanted dorsally in male Lewis rats and examined 2 and 4 weeks postoperatively. Microcomputer tomography (micro-CT) was used to examine the vessels formed. As shown in Fig. 7.12, examination of explanted hydrogels revealed approximately 6-fold and 12-fold increases in vascular densities at 2 and 4 weeks, respectively, for degradable gels compared to the densities in all other groups.

The effects of VEGF immobilized onto a porous collagen sponge for rapid vascularization were investigated in vivo [25]. Immobilization of the growth factor onto the sponge was carried out using 1-ethyl-3-[3-dimethylaminopropyl]carbodiimide hydrochloride (EDC) chemistry. The collagen patches were immersed in a solution of *N*-hydroxysulfosuccinimide (Sulfo-NHS) in phosphate-buffered saline (PBS) for 20 min. Then, the sponges were immersed in a solution of VEGF in PBS for 2 h, followed by washing in PBS. Patches with low (14.5 ± 1.4 ng) and high (97.2 ± 8.0 ng) amounts of immobilized VEGF, or without VEGF (control), were

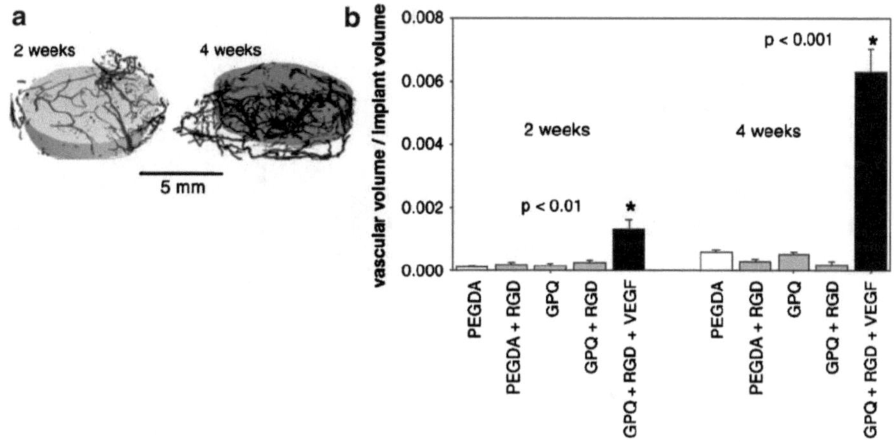

Fig. 7.12 Micro-CT images of bioartificial matrices implanted subcutaneously in rats and perfused with Microfil radiopaque contrast agent. (**a**) GPQ+RGD+VEGF implants showing vasculature in surrounding tissue growing into implant. The gray volume defines hydrogel. (**b**) Quantification of vascular volume/total implant volume. Data are represented as mean and SEM [24]

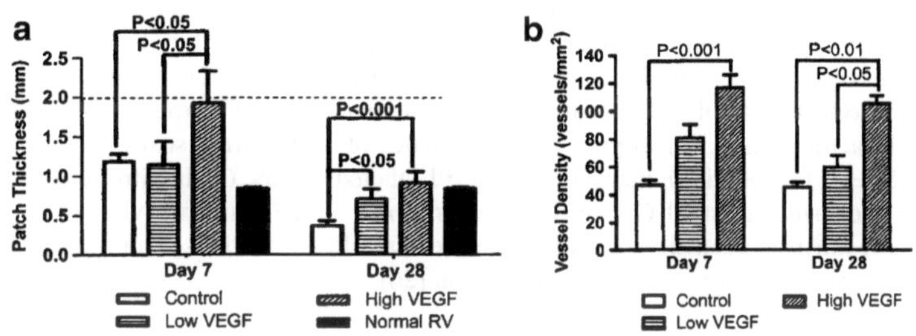

Fig. 7.13 (**a**) Patch thickness and (**b**) blood vessel density within the patches, at 7 and 28 days after implantation [25]

used to replace a full right ventricular free wall defect in rat hearts. Figure 7.13 shows the thickness of the patch (a) and blood vessel density inside the patch (b) at 7 and 28 days after implantation. The high-VEGF collagen sponge had greater blood vessel density ($p<0.01$) than the control at days 7 and 28 because of increased cell recruitment and proliferation ($p<0.05$ vs. control). At day 28, VEGF-treated sponges were significantly thicker ($p<0.05$) than the control, and the thickness correlated positively with neovascularization ($r=0.67$, $p=0.023$).

Multiple-growth-factor-delivery systems have been investigated. Richardson et al. studied the dual delivery of VEGF and PDGF for the formation of mature vessels and compared the results with those of the delivery of VEGF or PDGF alone [26]. Figure 7.14 represents the fabrication of a scaffold-containing growth factors and in vitro release kinetics of the factors. PDGF was pre-encapsulated in PLG microspheres prepared from polymers with two different intrinsic viscosities

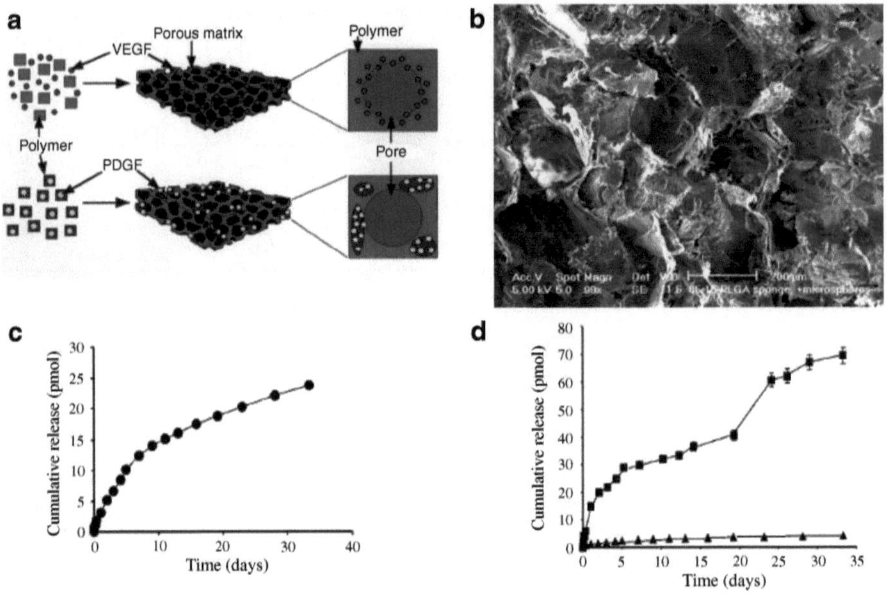

Fig. 7.14 Schematic of scaffold fabrication process and growth-factor-release kinetics. (**a**) Growth factors were incorporated into polymer scaffolds by either mixing with polymer particles before processing into scaffolds (VEGF) or pre-encapsulating the factor (PDGF) into polymer microspheres used to form scaffolds. (**b**) Scanning electron micrograph of a typical scaffold utilized for dual growth factor release. (**c**) In vitro release kinetics of VEGF from scaffolds fabricated from PLG (85:15, lactide:glycolide), measured using ^{125}I-labeled tracers. (**d**) In vitro release kinetics of PDGF pre-encapsulated in PLG microspheres (*triangle* 75:25, intrinsic viscosity= 0.69 dl/g; *square* 75:25, intrinsic viscosity=0.2 dl/g), before scaffold fabrication. Data represent the mean (n=5), and *error bars* represent standard deviation [26]

(0.69 and 0.2 dL/g) and processed by standard double emulsion. The scaffold was formed through an identical process using equal masses of PLG in microspheres, particulates with or without VEGF, and lyophilized alginate, using a high-pressure carbon dioxide fabrication process. ^{125}I-labeled growth factors were used for the in vitro release study. The scaffold released VEGF at the rate of 1.7 pmol/day (79 ng/day) for the first 7 days, followed by a lower delivery rate. The release rate of PDGF varied from 4.2 to 0.1 pmol/day depending on the MW of polymer used. The scaffold was implanted subcutaneously on the dorsal side of Lewis rats, and after 2 and 4 weeks, the implants were retrieved to investigate blood vessel formation from histological sections. Figure 7.15 represents the vascular density within tissue sections from each treatment condition. The bolus delivery of factors into blank scaffold was not sufficient for stable vessel formation, whereas the sustained delivery of growth factors rapidly formed dense vasculatures. In a hind-limb-ischemia model using nonobese diabetic (NOB) mice subjected to femoral ligation surgery, the dual delivery of VEGF and PDGF showed significantly higher vascular density than the treatment with single-growth-factor systems or the blank scaffold.

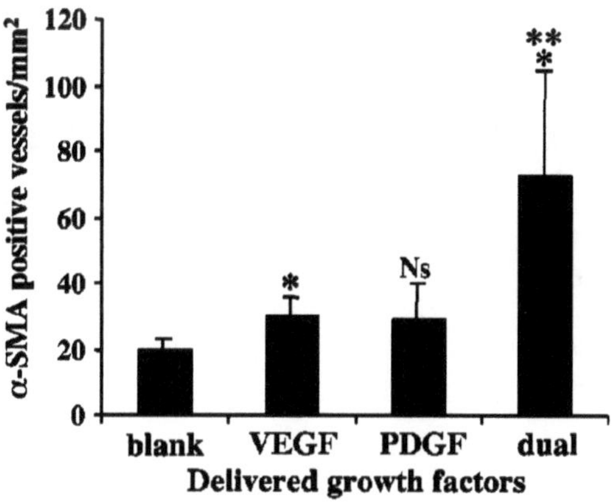

Fig. 7.15 Angiogenesis in nonobese diabetic (NOD) mice subjected to femoral ligation surgery. Sections of tissue adjacent to delivery scaffolds were stained for α-smooth muscle actin after 2 weeks (n=4) [26]

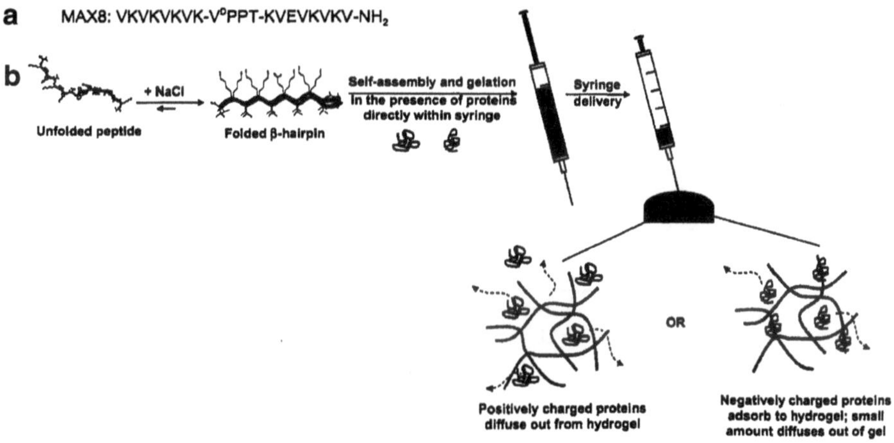

Fig. 7.16 (**a**) Sequence of MAX8. (**b**) Self-assembly of MAX8 into hydrogel directly in a syringe, subsequent delivery, and expected interactions between differently charged proteins and positively charged MAX8 hydrogel network [27]

An injectable peptide, which undergoes triggered hydrogelation in response to physiological conditions (pH 7.4, 150 mM salt, and 37°C) to form a mechanically rigid, viscoelastic gel with a net positive charge, was developed, and its efficacy as a protein-delivery system was evaluated [27]. MAX8 is a small, de novo designed β-hairpin peptide comprised of two β-strands of alternating valines and lysines with a glutamic acid at position 15 (Fig. 7.16a). A tetrapeptide (–V^{D}PPT-) sequence with high propensity to form a type II′ β-turn centrally connects the two strands. At low ionic strengths, MAX8 is freely soluble and remains unfolded due to the electrostatic repulsion between positively charged lysines. Increasing the ionic strength of

Table 7.4 Protein, molecular weight, hydrodynamic diameter, diffusion coefficient in water at 37°C, isoelectric point, and the net charge at pH 7.4 of model proteins [27]

Protein	Molecular weight (kDa)	Hydrodynamic diameter (nm)	D_{aq} (37°C) (10^{-8} cm^2/s)	Isoelectric point, pI	Charge at pH 7.4
α-Lactalbumin	14.1	3.2	142	4.2–4.5	–
Lysozyme	14.7	4.1	111	11	+
Myoglobin	17.4	3.9	113	7	Ø
Bovine serum albumin (BSA)	66	7.2	63	4.6–4.8	–
Lactoferrin	77	6.1	74	8.4–9.0	+
Human IgG	146	10.7	41	5.8–8.0	Ø

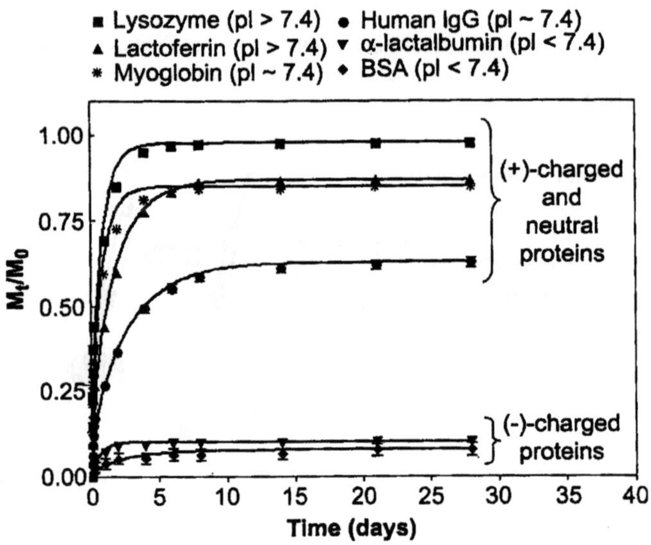

Fig. 7.17 Cumulative release profiles (Mt/M0) of proteins from 1 wt.% MAX8 hydrogels with 50 mM BTP, 150 mM NaCl, pH 7.4, at 37°C over a 1-month period [27]

the solution promotes the folding of the peptide into a facially amphiphilic β-hairpin structure (Fig. 7.16b). The N-terminal amine on each molecule of MAX8 contributes to the formation of a highly electropositive network. In the presence of proteins, hydrogelation can be triggered to directly encapsulate the macromolecules within the network. The release profiles of model proteins, such as lysozyme, α-lactalbumin, myoglobin, lactoferrin, BSA, and human immunoglobulin G (IgG), with various hydrodynamic diameters and surface charges, were examined from 1 wt.% MAX8 peptide hydrogels at 37°C. The properties of the proteins used are shown in Table 7.4, and their cumulative release profiles are shown in Fig. 7.17. Positively charged proteins were gradually released from the positively charged peptide hydrogel over several days. Lysozyme, a small protein with a net positive charge at pH 7.4, was released promptly. The release of negatively charged proteins, α-lactalbumin and BSA, was greatly hindered, and only 10% of these proteins were released at day 28.

To uniformly incorporate bFGF into a thick PGLA scaffold, heparin-conjugated PGLA (H-PGLA) was synthesized by a reaction between heparin and low-MW PGLA [28]. The H-PGLA was then mixed with high-MW PGLA to create scaffolds. The binding efficacy of bFGF increased with increasing H-PGLA content. The incorporated bFGF was released in vitro slowly over 2 weeks, with the bioactivity of bound bFGF being preserved. The H-PGLA/PGLA scaffold showed better attachment, growth, and cell affinity in mouse 3T3 cells than the PGLA scaffold.

Wound Healing by Platelet-Rich Plasma

A single growth factor has been applied in most studies on wound healing and tissue regeneration, although the naturally occurring tissue regeneration process must be regulated by a set of multiple growth factors at a determined time schedule and concentration, as is seen generally in wound healing. The cost of growth factors can also be quite expensive.

Platelet-rich plasma (PRP) contains a wide variety of growth factors. PRP is defined as autologous, concentrated platelets in a small volume of plasma [29]. The advantages of using PRP are that it is made from the patient's own plasma and hence is completely safe, relatively inexpensive, and expected to release the right mixture of growth factors for wound healing. There are many ways to name PRP, such as concentrated PRP (cPRP), plasma rich in growth factors (PRGF), platelet-enriched fibrin glue, platelet-rich fibrin (PRF), or platelet-rich fibrin matrix (PRFM). Various methods to synthesize PRP are available, resulting in different compositions (number of platelets, amount of fibrin, and with or without leukocytes) [30]. The effective concentration of platelets in PRP is thought to be around 1,000,000/μL, or 4–7 times of the normal blood [29].

PRP has long been clinically used in the areas of oral and maxillofacial surgery, orthopedics, and plastic surgery. Many studies demonstrate the positive effects of PRP in either bone or soft-tissue healing, although several other research groups reported little or no benefit from PRP. A possible reason is the bolus release of growth factors from PRP; sustained release of growth factors may be preferable.

Platelets

Blood consists of a liquid component (plasma) and blood cells (red blood cells [RBCs], white blood cells [WBCs], and platelets). The typical weight proportions of blood components are shown in Table 7.5.

Platelets are small discoid blood cells produced in bone marrow. They are oval in shape, ~2 μm in diameter, and their life span is 7–10 days. Platelets have a trilaminar

Table 7.5 Composition of human blood

Blood	Blood cells 40–50% (v)	Red blood cell 96% (w)	
		White blood cell 3% (w)	Granulocyte 65% (w)
			Monocyte 5% (w)
			Lymphocyte 30% (w)
		Platelets 1% (w)	
	Plasma 50–60% (v)	Water 90% (w)	
		Proteins 7% (w)	Albumin 57% (w)
			Fibrinogen 4% (w)
			Immunoglobulin 15% (w)
			Others 24% (w)
		Others 3% (w) (fats, inorganic salts, etc.)	

Table 7.6 Major growth factors available in α-granules of platelet

Growth factors	Abbreviation
Platelet-derived growth factor	PDGF
Transforming growth factor alpha and beta	TGF-αβ
Epidermal growth factor	EGF
Fibroblast growth factor	FGF
Insulin growth factor	IGF
Platelet-derived epidermal growth factor	PDEGF
Platelet-derived angiogenesis factor	PDAF
Interleukin-8	IL-8
Tumor necrosis factor-alpha	TNF-α
Connective tissue growth factor	CTGR
Granulocyte macrophage colony-stimulating factor	GM-CSF
Keratinocyte growth factor	KGF

cell membrane with a glycoprotein receptor surface overlying, and partially interspersed with and penetrating, a bilayer of phospholipid and cholesterol. Platelets lack nuclei but contain organelles and structures such as mitochondria, microtubules, and granules. There are approximately 50–80 α-granules per platelet, which contain over 30 types of bioactive proteins, many of which have a fundamental role in hemostasis and tissue healing (Table 7.6). When activated, platelets change shape, develop large, sticky protuberances or pseudopodia, and allow α-granules to diffuse out and release their protein contents to surroundings [31] (Fig. 7.18).

Preparation of PRP

In general, PRP is prepared by centrifugation of whole blood to remove RBCs and plasma and to concentrate platelets [32] (Fig. 7.19). Blood is collected in the presence of an anticoagulant (usually citrate ions, such as Anticoagulant Citrate Dextrose-A; ACD-A). An initial centrifugation at low speed separates the blood

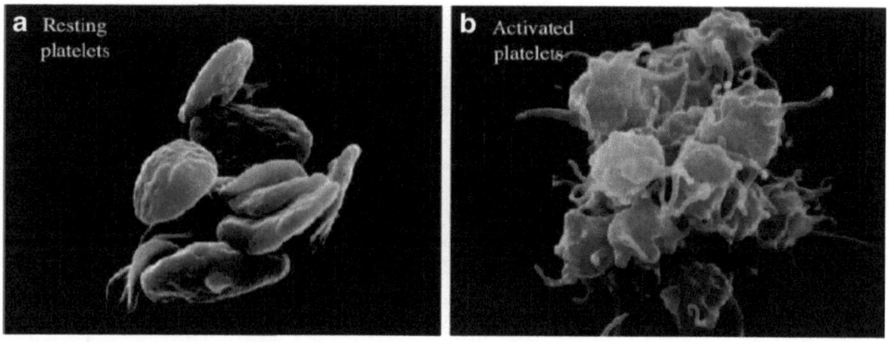

Fig. 7.18 Mode of action of PRP [31]

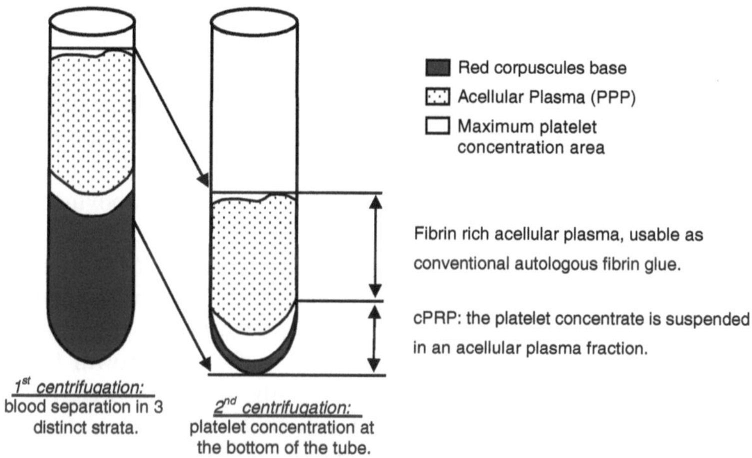

Fig. 7.19 Technology of cPRP processing [32]

into three layers as a function of density: the top plasma layer, the middle buffy coat consisting of WBCs and platelets, and the bottom RBC layer. The top and middle layers are then centrifuged at high speed to separate platelet-poor plasma (PPP) and platelet-rich plasma (PRP). The volume of PRP collected is approximately 10% of the total blood used.

Automatic separation devices are commercially available, including the Gravitational Platelet Separation system (GPS; Biomet, Warsaw, IN.), Platelet Concentrate Collection System (PCCS; Implant Innovations, Inc., Palm Beach Gardens, FL.), Symphony II (DePuy, Warsaw, IN.), SmartPReP (Harvest Technologies Corp., Norwell, MA.), and Magellan (Medtronic, Minneapolis, MN.) [33]. In addition, standard cell separators and salvage devices can be used to produce platelet concentrates. They include CATS (Fresenius, Wilmington, DE.), Sequestra (Medtronic, Minneapolis, MN.), and Haemonetics Cell Saver 5 (Haemonetics Corp., Braintree, MA.) [33].

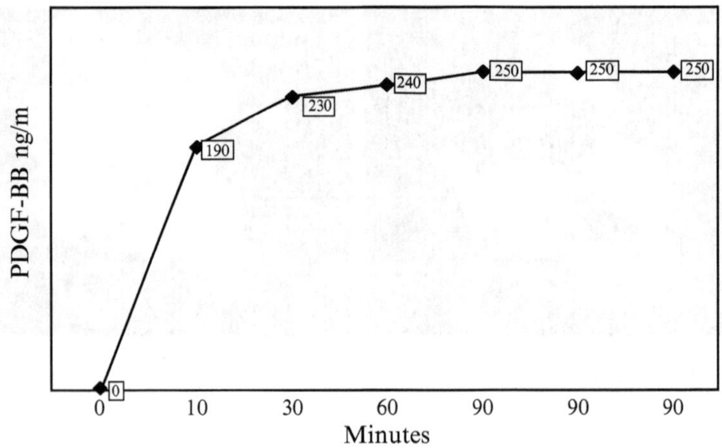

Fig. 7.20 Release kinetics of PDGF-BB from activated PRP [35]

Before application, PRP must be activated to release the growth factors from the α-granules, and clotted to allow for their delivery. The most common method of platelet activation for clinical use is to add calcium chloride and thrombin. Typically, bovine thrombin is used in commercial products; however, use of bovine thrombin can lead to complications associated with the formation of antibodies against the bovine thrombin, a rare but serious complication [34]. Therefore, autologous thrombin has been used in most studies. PRP can be stored for 8 h at room temperature and is activated immediately before use, since the growth factors are released soon after the activation of platelets. According to Marx, 70% of growth factors are secreted within 10 min and close to 100% within the first hour (Fig. 7.20) [35]. Other platelet activation methods include freeze/thaw, collagen, and adenosine diphosphate (ADP). Fusa et al. compared the clotting of PRP using bovine thrombin or collagen [36]. The clot formed on contact with collagen was less retracted, but indicated equal release of growth factors.

The functional importance of PRP components other than platelets, i.e., leukocytes and fibrin, has been recognized. Leukocytes produce high amounts of VEGF and affect angiogenesis, especially during bone regeneration. They also play an important role in combating infection and mounting an immune reaction. The fibrin gel formed during PRP processing can act as a scaffold as well as a carrier for growth factors. Attempts have been made to form a denser and more and rigid fibrin network.

In a review, Dohan et al. classified platelet concentrates into four different groups according to their composition: pure platelet-rich plasma (P-PRP), leukocyte- and platelet-rich plasma (L-PRP), pure platelet-rich fibrin (P-PRF), and leukocyte- and platelet-rich fibrin (L-PRF) [30]. In the preparation of Choukroun's PRF, blood is collected without any anticoagulant and immediately centrifuged. This results in clot formation during the centrifugation step, and L-PRF is obtained (Fig. 7.21) [32]. This is perhaps the simplest and most inexpensive protocol for the production

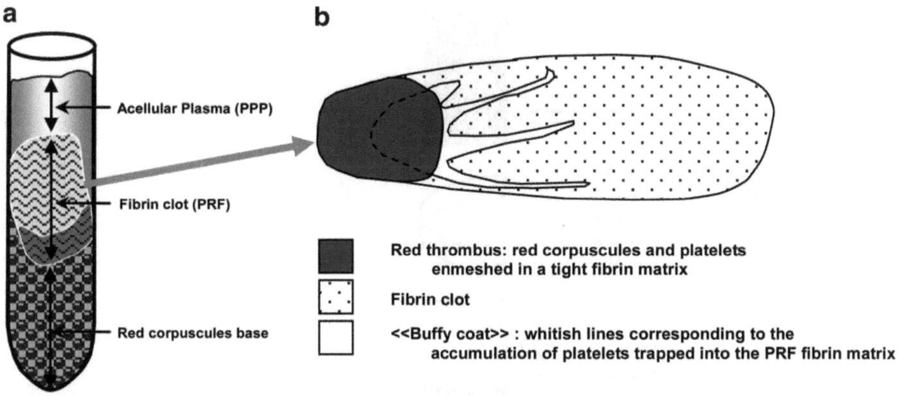

Fig. 7.21 (**a**) PRF formation in the middle of the tube after centrifugation immediately after blood collection; (**b**) PRF is divided into three parts: a red thrombus in contact with the red blood corpuscle base, an acellular fibrin gel, and a network of buffy columns corresponding to platelet accumulation [32]

of a sustained carrier of growth factors, since this eliminates the use of anticoagulant, requires just a single centrifugation step, and uses natural fibrin clot polymerization. PRF will be described in more detail later.

To increase the concentration of fibrinogen to form a rigid gel, Thorn et al. studied a method for the preparation of strong autologous fibrin glue with platelet growth factors [37]. The preparation is schematically represented in Fig. 7.22. Two hundred milliliters of patients' blood was collected for glue preparation. Fibrinogen solution enriched with growth factors (FS-GF) was produced from the precipitation of PRP in cold ethanol. Thrombin was prepared from the dilute plasma by precipitation of euglobulins at low pH. The precipitate containing prothrombin, fibrinogen, and other coagulation factors was redissolved in $CaCl_2$ solution, following which the pH was adjusted to neutral to start clot formation. The thrombin solution was immediately removed from the clot. This technique was performed in a closed system within 60–90 min after blood collection. The fibrinogen concentration prepared from this system was 12 times higher than that of PRP (32 and 2.6 mg/mL for FS-GF and PRP, respectively), and the PDGF concentration was eight times higher (243 and 29 ng/mL for FS-GF and PRP, respectively). This glue was clinically successful when used with particulate cancellous bone in reconstructive maxillofacial surgery.

Growth Factor Concentrations

Depending on the individuals and the activation conditions used (i.e., amounts of thrombin and calcium), the amount of growth factors released from PRP can be quite different [38]. Epply et al. analyzed growth factor concentrations of PRP from

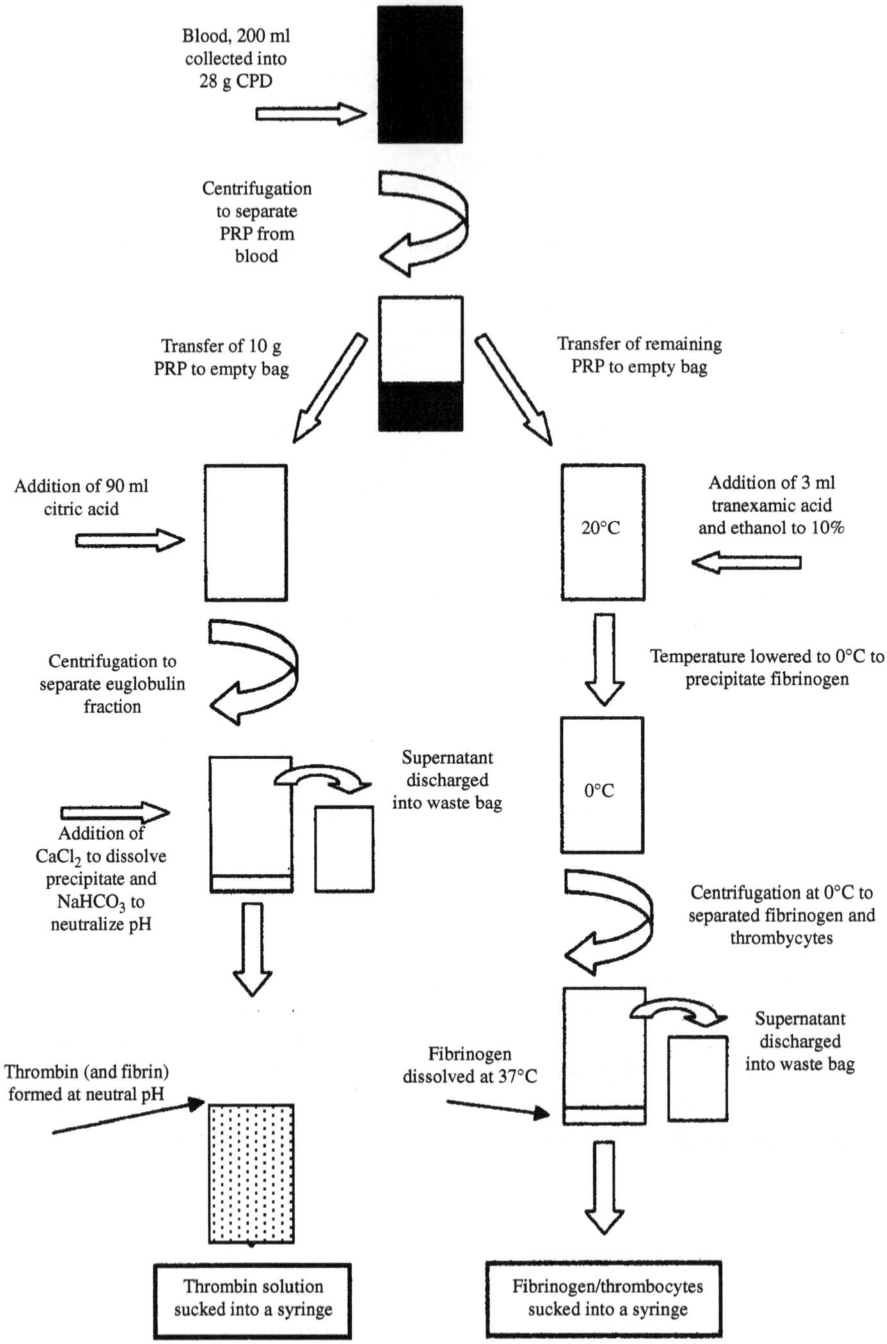

Fig. 7.22 Flowchart for the preparation of autologous fibrin glue with growth factors. *CPD* citrate phosphate dextrose, *thrombocyte* = platelet [37]

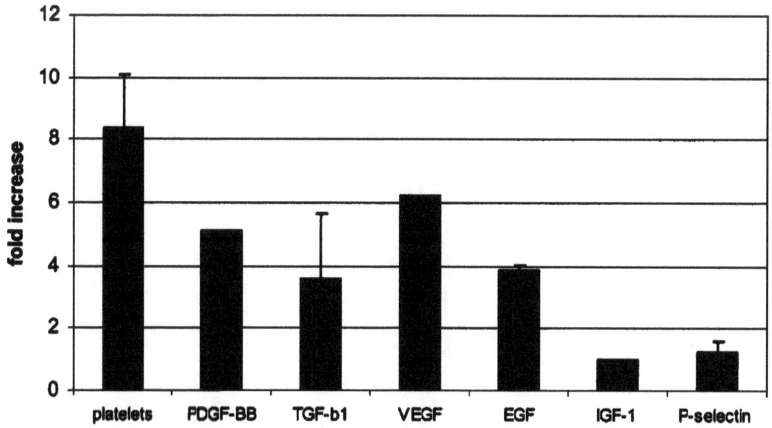

Fig. 7.23 Fold increase in platelet and growth factor concentrations, calculated from averaged means of concentrations from patients studied ($n = 10$) [33]

ten healthy patients undergoing cosmetic surgery [33]. PRP was produced by centrifugation using the GPS system, and the average platelet concentration was 1,603,000 ± 330,000/μL. PRP was activated with 1,000 units of bovine thrombin and $CaCl_2$. The concentration of growth factors in PRP was measured using ELISA, and the averages of these concentrations were 17 ± 8 ng/mL for PDGF-BB (an isoform of PDGF), 120 ± 42 ng/mL for TGF-β1, 955 ± 1,030 pg/mL for VEGF, 470 ± 317 pg/mL for EGF, and 72 ± 25 pg/mL for insulin-like growth factor (IGF-1). The increase in average concentrations of growth factors in PRP compared to the concentrations in whole blood is given in Fig. 7.23.

The influence of gender and age on the growth factor content of PRP was investigated by Weibrich et al. They collected 115 specimens of PRP and whole blood from healthy men and women aged 17–62 years [39]. PRP was prepared by discontinuous cell separation, and the platelet concentration of PRP was found to be 1,407,640 ± 320,100/μL, five times higher than that of WB (266,040 ± 60,530/μL). After one freeze/thaw cycle and centrifugation, the concentration of PDGF-AB, PDGF-BB, TGF-β1, TGF-β2, and IGF-I in the supernatants was measured with ELISA. The data are shown in Fig. 7.24. No correlation was found between growth factor content and platelet count. There was also no effect of patient gender and age on platelet count or growth factor content of PRP, except for IGF-I.

PRP produced using two different automatic systems, the Curasan-type kit and the PCCS system, were compared [40]. Platelet counts were higher in PCCS PRP (2,232,500/μL) than in Curasan (1,140,500/μL), consistent with higher levels of TGF-β1.

The release of growth factors (i.e., TGF-β1, PDGF-AB, and VEGF) from PRP activated using either thrombin or collagen was investigated by Harrison et al. [41]. PRP was prepared using HarvestSmartPreP2 System and clotted using either bovine thrombin in 10% calcium chloride solution or a collagen sponge prepared by

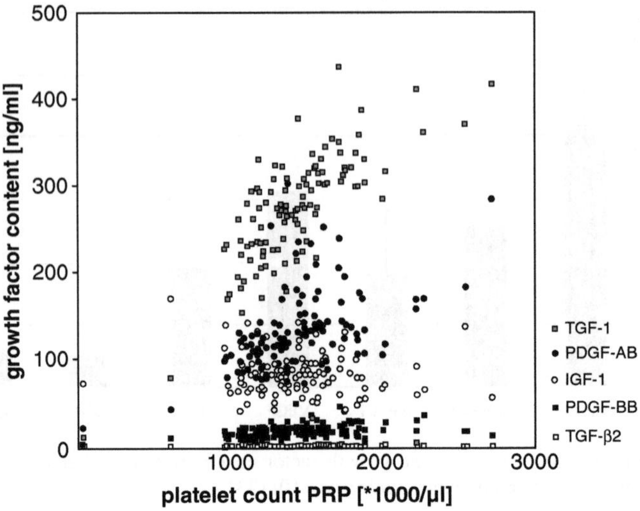

Fig. 7.24 Scatter plots of thrombocyte count versus PDGF-AB, PDGF-BB, TGF-β1, TGF-β2, and IGF-I levels in PRP samples [39]

lyophilizing atelocollagen solution. Whole blood was clotted using thrombin or without using anticoagulant. TGF-β1 and PDGF-AB were immediately released from PRP upon activation with bovine thrombin, whereas collagen-activated PRP slowly released these factors. The cumulative amount of TGF-β1 from collagen-activated PRP was 80% greater than that of thrombin-activated PRP over 7 days. The release profile of VEGF was not affected significantly by activator choice and was incremental over the 7-day period. It is suggested that the gradual release of presynthesized VEGF occurs from WBC granules that are concentrated fivefold in PRP compared to in whole blood, and hence, the type of platelet activator used does not affect its release.

In Vitro Studies

Effects of PRP extracts on several cell types were investigated by Kakudo et al. They evaluated the effects of PRP on human adipose-derived stem cells and human dermal fibroblasts [42]. The cells were cultured with serum-free Dulbecco's Modified Eagle Medium (DMEM) supplemented with 1%, 5%, 10%, and 20% activated platelet-rich plasma for 7 days. PRP (containing a 7.9-fold concentration of platelets) was prepared using a double-centrifugation process and activated with calcium and autologous thrombin. The proliferation results show that addition of 5% activated PRP to the culture medium promoted cell proliferation to the highest extent, while 20% PRP did not (Fig. 7.25).

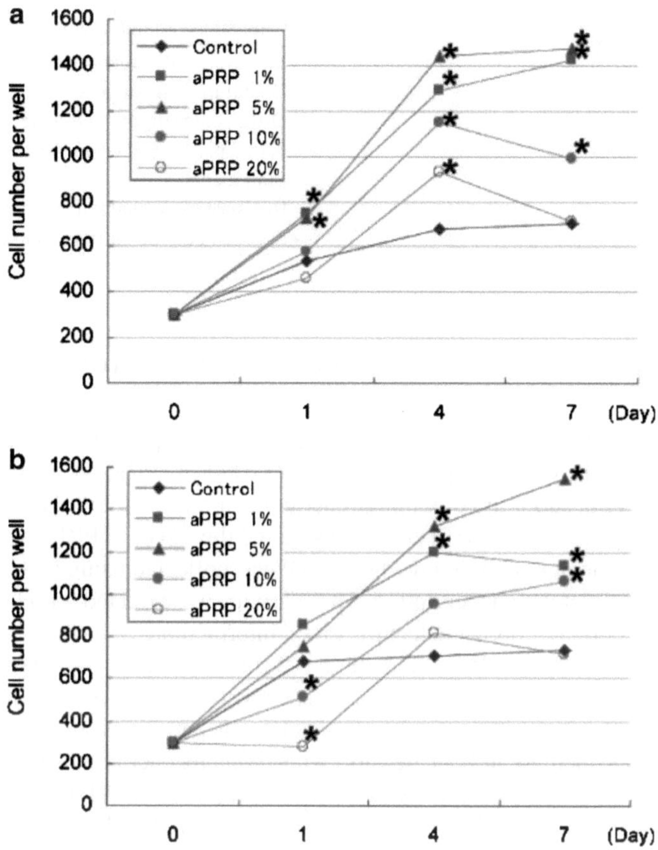

Fig. 7.25 Proliferation of human adipose-derived stem cells (**a**) and human dermal fibroblasts (**b**) on days 1, 4, and 7 after the addition of various concentrations of activated platelet-rich plasma (*aPRP*) [42]

Graziani et al. investigated the effect of different concentrations of activated PRP on oral fibroblast and osteoblast functions [43]. The maximum effect of activated PRP was achieved at a 2.5-fold concentration of platelets in PRP, with higher concentrations resulting in reduction of cell proliferation. Choi et al. examined the influence of PRP concentrations on the viability and proliferation of alveolar bone cells [44]. Cell viability and proliferation were suppressed with high concentrations of PRP, but stimulated with low PRP concentrations (1 and 5%). Liu et al. suggested that the proliferation of fibroblasts exposed to concentrated platelets was pH dependent [45]. Under acidic conditions, platelets were activated and PDGF was released from the α-granules of activated platelets, leading to fibroblast proliferation. However, the concentration of TGF-β was lower at pH 5.0 than at both pH 7.1 and 7.6.

Table 7.7 Autologous PRP-clots as an aid to healing [46]

Clinical use of autologous PRP-clots
Maxillofacial surgery and bone grafts
Dental implant surgery
Orthopedic surgery and bone reconstruction
Facial plastic and cosmetic surgery
Skin ulcers
Eye surgery-retinal hole repair
Sports medicine-cartilage and tendon repair

Animal and Clinical Studies (Bone and Soft Tissues)

Clinical areas of PRP applications are shown in Table 7.7 [46]. The effective platelet concentration seems to be approximately 1,000,000/μL PRP. At lower concentrations, the effect is suboptimal, while higher concentrations might have a paradoxically inhibitory effect. Weibrich et al. investigated the effect of PRP on peri-implant bone regeneration in a rabbit model [47]. PRP was collected using PCCS to analyze the influence of platelet concentration. Three groups were treated with low (0.5–1.5-fold, i.e., 164,000–373,000 platelets/μL PRP), intermediate (2–6-fold, i.e., 503,000–1,729,000 platelets/μL PRP), and high (9–11-fold, i.e., 1,845,000–3,200,000 platelets/μL PRP) platelet concentrations, respectively. After 4 weeks of implantation, only the group treated with intermediate platelet concentrations showed a significant difference (90% increase in bone density) when bone regeneration was compared.

The effect of platelet-derived wound-healing factors (PDWHF) on accelerating angiogenesis in wound healing was evaluated in a rat model [48]. Collagen sponges (1-cm diameter and 2–3-mm thick) were saturated with 300 μL of PDWHF, which was prepared from 5×10^9 platelets, and implanted subcutaneously in rats. Six hours to 14 days postimplantation, vascular corrosion casts were prepared and examined with scanning electron microscopy (SEM). It was found that leukocyte migration and angiogenesis were accelerated at 24–48 h, and the number of vessels formed in the PDWHF-collagen sponge group was greater than that of the control group (with the collagen sponge soaked in buffer). Using the Evans blue perfusion method, the quantity of vessels in sponges after 14 days postoperation was evaluated. It was found that the number of vessels formed in the PDWHF group was twice that in the control group. Since the amounts of TGF-β and PDGF-BB in PDWHF were significantly less than those required to achieve angiogenic and wound-healing responses, it suggested that a combination of TGF-β, PDGF-BB, and other components which are not yet defined is effective for angiogenesis [26].

Cho et al. investigated the effect of PRP on wrinkle reduction on ultraviolet B (UVB)-induced skin wrinkles in nude mice [49]. Mice were photoaged by UVB irradiation for 8 weeks, and PRP or saline was injected, while the control group had no treatment. After 4 weeks, the degree of wrinkle formation was evaluated; the reduction was more significant in the PRP-injected group than in the other groups. The dermal layer was greatly thicker in the PRP-injected group.

Fat grafting has been used extensively as a soft-tissue filler, but survival of fat graft is inconsistent. There is no definitive method for fat processing that ensures maximal graft take and viability. Therefore, the effect of PRP on human fat graft survival was evaluated in nude mice [50]. PRP was prepared using double centrifugation. The mixture of fat (0.7 mL) and PRP (0.21 mL) was activated with thrombin/calcium chloride solution (0.15 mL) and injected subcutaneously in mice using a 16-G needle. Ten weeks after implantation, the volume and weight of fat grafts were found to be significantly higher in the PRP group than in the control group (fat in saline). Cellular integrity and degree of inflammation were not statistically different between the two groups. Hence, PRP treatment was concluded to improve the survival of fat grafts.

Zechner et al. used a miniature pig model to investigate the effect of PRP on bone regeneration [51]. They found that bone regeneration was roughly double that of control after 6 weeks of implantation, when PRP containing 800,000–1,000,000 platelets/μL was used. Using a dog peri-implant model, Kim et al. found an increase in bone regeneration at a bone defect (bone density; 43 ± 15.4% vs. 72 ± 16.4%) [52]. In clinical studies, Marx et al. showed an increase in bone regeneration after 6 months using PRP containing 595,000–1,100,000 platelets/μL (bone density; 55.1 ± 8% vs. 74 ± 11%) [53].

Cervelli et al. [54] mixed PRP with fat grafting and used it for 43 patients who underwent plastic, reconstructive, or maxillofacial surgery for chronic lower extremity ulcers ($n = 18$) and multiple facial applications ($n = 25$). As shown in Fig. 7.26, 88.9% patients had 100% reepithelization during a 9.7-week course of twice-daily wound treatment with PRP suspended on a collagen base, compared with 60% for the control group treated with hyaluronic acid and collagen medication. For patients with facial disease who underwent facial three-dimensional reconstruction by fat grafting and PRP, 70% maintenance of contour restoration and three-dimensional volume was achieved after 1 year, compared with only 31% in control patients treated with only fat grafting. They also demonstrated that PRP increased the growth of adipose tissue by stem cells in vitro to a significantly greater extent than the control.

The efficacy of PRP on wound healing of chronic diabetic foot ulcers was studied in comparison with PPP [55]. Twenty-four patients were randomly divided into two groups: PRP group ($n = 12$) and PPP group ($n = 12$). PRP and PPP were prepared by double centrifugation and activated with thrombin/$CaCl_2$ solution. The average ulcer area was 4–21 cm^2. PRP or PPP was applied to the ulcer, followed by Vaseline gauze and then a dressing. The dressing was changed twice weekly for up to 20 weeks or till healing occurred. There was no statistically significant difference between the two groups regarding clinical signs. The mean healing times for PRP and PPP were 11.5 and 17 weeks, respectively. This result suggests that release of growth factors from PRP accelerated the healing of chronic diabetic ulcers. Another study using platelet gel in combination with autologous fibrin glue for skin grafting on chronic lower extremity ulcers showed enhanced wound healing [56].

Messora et al. [57] investigated bone healing of critically sized defects (8-mm diameter, Fig. 7.27) in rat calvaria treated with PRP activated with two different methods: $CaCl_2$ and thromboplastin solutions. As seen in Fig. 7.28, the group treated

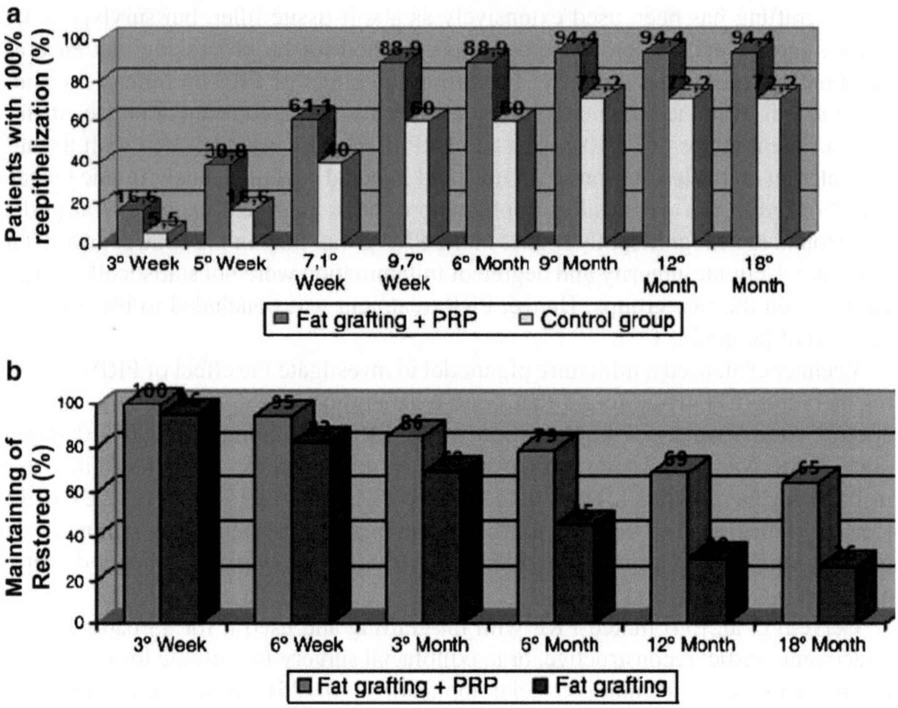

Fig. 7.26 Bar graphs showing effects of PRP on (**a**) skin chronic ulcer reepithelization compared with the effects on the control group and (**b**) maintenance of restored fat compared to that in the control groups [54]

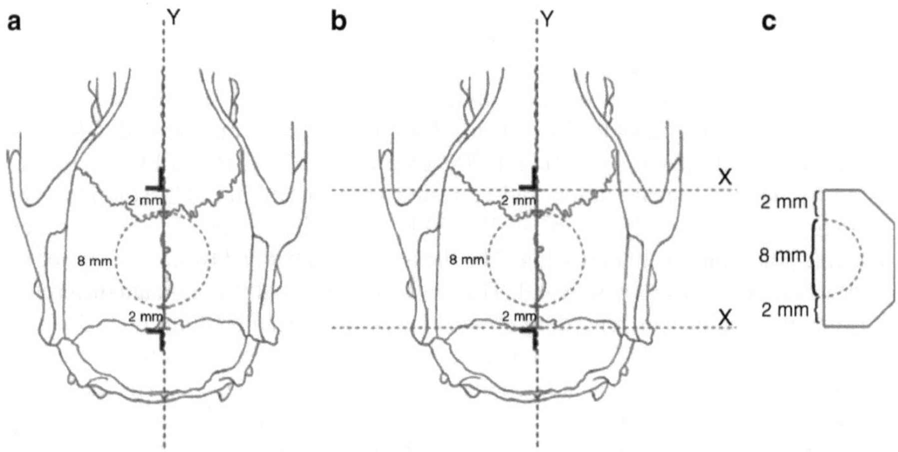

Fig. 7.27 (**a**) Longitudinal cut along the center line (Y) of a critically sized defect, (**b**) transverse cuts (X), and (**c**) dimensions of specimen to be embedded in paraffin [57]

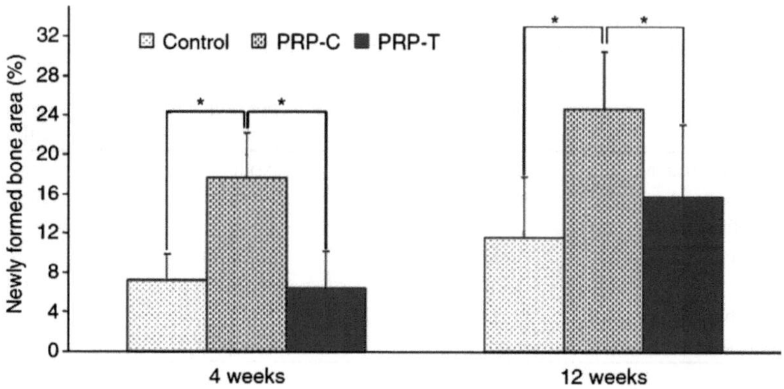

Fig. 7.28 Means and standard deviations of newly formed bone area (expressed as% of total defect area) for Control, $CaCl_2$-activated PRP (PRP-C), and thromboplastin-activated PRP (PRP-T) groups, 4 and 12 weeks postoperatively. *Statistical difference between the groups ($p<0.05$) [57]

with $CaCl_2$-activated PRP showed a statistically greater amount of bone formation than the control or the thromboplastin-activated PRP group. The authors hypothesized that different activators used to initiate PRP clot formation influence growth factor levels, their distribution, and release kinetics.

For maxillary sinus augmentation with the simultaneous placement of dental implants, Rodriguez et al. combined the use of PRP with deproteinated bone xenograft [58]. They assumed that PRP would lead to improved revascularization and faster consolidation of the graft through its angiogenic properties when combined with an osteoconductive material. Fifteen patients with residual alveolar height <5 mm in the posterior maxillary alveolus underwent a total of 24 maxillary sinus augmentations. Before surgery, PRP was obtained from the required amount of autologous blood and then mixed with the bone xenograft. Dental implants were inserted simultaneously in the grafted sinuses. If there were perforations in the sinus mucosal membrane, these were approximated and covered with autologous fibrin glue prepared from the centrifuged serum with bovine thrombin and calcium chloride. Following this, the bovine graft mixed with PRP was placed in the maxillary sinus surrounding the dental implants. The bone biopsy from the patients showed evidence of viable new bone formation in close approximation to the xenograft. The density of the grafted bone was similar to or exceeded the density of the surrounding native maxillary bone.

Choi et al. conducted a histological examination of grafted bone at 6 weeks postoperatively to evaluate new bone formation during the early phase, and both sides of the dog mandible were used to provide matched pairs [59]. The mandibular premolar teeth had been bilaterally extracted previously, and the ridges were allowed to heal for 3 months. After this period, continuity resection was performed on both sides of the mandible. One defect (the PRP group) was reconstructed with the original particulate bone mixed with PRP. As a control, the contralateral defect (non-PRP group) was reconstructed with the original particulate bone alone. Biopsies after

Table 7.8 Comparison of products of autologous platelet gel and fibrin glue prepared by concentrate system, autologous platelet gel prepared by autotransfusor, and tisseel fibrin sealant [60]

	Autologous platelet gel and fibrin glue prepared by concentrate system	Autologous platelet gel prepared by autotransfusor	Tisseel fibrin sealant
Risk of disease transmission	None	None	Low
Platelet-associated growth factors	Present	Present	Absent
Fibrinogen concentration	Moderate	Moderate	High
Cost	~$350	~$500	~$850
Amount of blood necessary	90–180 cc	500 cc	0
Convenience/ease of use	Easy	Complex	Very easy

6 weeks showed lower levels of bone formation in the PRP group than in the non-PRP group, and fluorescence microscopy revealed a delay in the remodeling of grafts loaded with PRP. These findings suggest that the addition of PRP does not appear to enhance new bone formation in autologous bone grafts.

Man et al. evaluated the efficacy of PRP and PPP in cessation of capillary bleeding in the surgical flaps of patients undergoing cosmetic surgery [60]. Twenty patients were involved in this study and underwent surgical procedures including face-lifts, breast augmentations, breast reductions, or neck lifts. PRP and PPP were prepared from autologous blood (90–180 mL) using an automated autologous platelet concentration system (SmartPReP). The PPP and PRP were combined with a thrombin-calcium chloride solution to produce autologous fibrin glue and platelet gel, respectively. For application, two syringes were prepared, one containing 10 mL of PRP or PPP and the other containing 1 mL of thrombin-calcium chloride solution (1,000 units/mL of thrombin in 10% calcium chloride). They were connected to a dual spray applicator tip which allowed both solutions to be mixed as they were administered onto the injury site. The autologous gels were formed in 15–45 s, and the oozing was effectively stopped. Compared to commercial fibrin products, PRP and PPP were cost-effective and were not hampered by considerations of disease transmission (Table 7.8).

Slow Release of Growth Factors

Currently available forms of autologous PRP are usually fragile, unstable, and prone to rapid fibrinolysis and dissolution following implantation. They are not easy to handle in a clinical setting and too weak as scaffolds when used in association with isolated cells.

Few studies have been published to examine whether growth factors released from degranulating platelets remain at the site of action for the desired time period, or diffuse away within a short time. Once PRP is activated and the clot formed, the clot starts to retract, and the secretory proteins leach out. Hence, the clotted PRP

Table 7.9 Comparison between total amounts of various growth factors released after 168 h and initial amounts extracted from Choukroun's PRF membrane [63]

Tested molecules	TGF-β1 (ng)	VEGF (pg)	PDGF-AB (ng)	TSP-1 (μg)
Total released after 168 h (sum of the amounts measured at each experimental time)	273.4±15.3	6,071±773	50.3±6.3	26.1±1.93
Total extracted from the membrane at T_0	39.2±2.8	953±168	32±3.5	2.8±0.18
Ratio between slow release and extracted molecules	6.97	6.37	1.57	9.3

should be transferred to the application site within 10 min of clot initiation [29]. It has been found that close to 100% of the growth factors are released from the commonly used PRP within the first hour after activation [35]. This might explain the clinical reports of acceleration of healing in the short term, but without significant mid- or long-term effects [61]. Therefore, efforts have been made to create slow-release systems; among these is PRF.

PRF was originally introduced by Choukroun et al. in 1998 specifically for use in oral and maxillofacial surgery. Clinical studies performed in implant installation in sinus lifting and alveolar bone defects [32, 62] revealed enhancement of bone formation and an increased healing rate of grafted bone. Combination of PRF and autogenous bone increased the rate of osteogenesis and the quality of new bone.

In vitro release of growth factors (TGF-1, PDGF-AB, VEGF) and a coagulation matrix glycoprotein (thrombospondin-1, TSP-1) from PRF was investigated [63]. After centrifugation of whole blood, PRF was obtained and gently pressed into membrane shape before placing in 4 mL of sterile DMEM. Slow release of these proteins was observed at least for 1 week, as shown in Table 7.9. Massive release was found for TSP-1, a coagulation clot reinforcement agent, which could work as a natural antihemorrhagic agent at a surgical site.

Kang et al. fabricated Choukroun's PRF and obtained its extract (PRFe) after freezing at −80°C [64]. The properties of Choukroun's PRF were compared with those of concentrated PRP. Matrix metalloproteinase 9 (MMP-9) and Serpin E1 were detected from PRFe but not in concentrated PRP. The presence of MMP-9 in PRF was expected, as it is one of the key factors for differentiation of human alveolar bone marrow stem cells (hABMSCs) to mineralization. Fresh PRF was transplanted into a critically sized defect of mouse calvaria. Six weeks after implantation, bone regeneration, covering most of the defect, was better in the PRF group, than in the control group.

Human dental pulp cells (DPCs) were cultured in vitro on Choukroun's PRF and pressed into membranes [65]. PRF was found to stimulate the proliferation and differentiation of DPCs during 5-day incubation, upregulating osteoprotegerin and ALP activity.

The efficacy of PRF as a scaffold for periosteal tissue engineering was investigated in vitro and compared with the commonly used collagen membrane [66]. PRF was prepared by immediate centrifugation of blood without anticoagulant for 12 min at 2,700 rpm. A PRF membrane was obtained by squeezing the serum out from the clot. Human periosteal cells were cultured on the membrane piece to assess cell vitality. Both quantitative and qualitative examinations revealed higher values for PRF, and the PRF membrane was concluded to be a suitable scaffold for bone tissue engineering.

The effectiveness of PRF in clinical applications has been reported by several groups. PRF was implanted as the sole-filling material during simultaneous sinus lifting in 25 sinus elevations on 20 patients [62]. After 6 months, panoramic X-ray and three-dimensional volumetric computed radiography (VCR) were performed to evaluate the subsinus residual bone height and the final bone gain around the implants. Significant bone gain was observed for all implants, indicating implant stabilization with a high volume of natural regenerated bone in the subsinus cavity up to the tip of the implants. The biopsies of nine patients showed well-organized, vital bone.

Lucarelli et al. analyzed platelet-rich fibrin matrix (PRFM) obtained using a commercially available system (FIBRINET®) [67]. PRFM demonstrated robust mechanical properties with a tear elastic modulus of 937.3 ± 314.6 kPa, stress at break of 1,476.0 ± 526.3 kPa, and elongation of 146.3 ± 33.8%. Microscopic observation showed that platelets were located on one side of the matrix, while the average dimension of fibrin fibers was augmented more strongly than the native fibrin gel. The fibrin-based clots had an elastic modulus or stiffness in the range of 0.1–500 Pa, depending on the clot structure and physiologic conditions. In contrast, PRFM made of the same components had an elastic modulus of 937.3 kPa, which is comparable with that of arterial tissue, approximately 50% of the stiffness of intact human skin, and 600 times stiffer than the fibrin clot. The release kinetics of PDGF-AA, PDGF-AB, TGF-β1, VEGF, EGF, and bFGF from PRFM are shown in Fig. 7.29, where the initial burst of growth factors is observed. Figure 7.30 shows mesenchymal stem cells (MSC) proliferation induced by PRFM-conditioned media (CM). Cell proliferation was determined using the methylene blue assay after 24, 48, and 72 h in culture. CM was obtained from the PRFM of three donors. CM with concentrations up to 20% was effective for MSC proliferation.

Different materials and methods have been investigated to achieve slow release of PRP growth factors for bone regeneration. Tsay et al. investigated the effect of thrombin receptor activator peptide-6 (TRAP; sequence: SFLLRN) as activator for PRP [68]. PRP was clotted using bovine thrombin, TRAP, or bone substitutes (Allogro, Ceramed, Lakewood, CO; BioGlass, Mo-Sci, Rolla, MN; or BioOss, Osteohealth, Shirley, NY). The release of PDGF and TGF-β was quantified using ELISA. TRAP is a synthetic peptide corresponding to an amino-terminal peptide sequence that mimics thrombin in eliciting thrombin-signaled cell responses in platelets, independent of receptor cleavage [69, 70]. Figure 7.31 shows release profiles of PDGF and TGF-β from PRP clotted with different materials [68]. The TRAP and bone substitute groups released significantly less PDGF and TGF-β than the bovine thrombin group at day 1, and slow release of these factors was observed during 14 days.

Two types of alginate hydrogel-based PRP delivery systems (beads and capsules) were investigated for bone regeneration [71]. The release profiles of growth factors (PDGF, TGF-β1, and IGF) and the effects of these factors on the growth and ALP activity of human osteoblast-like cells were evaluated. Beads were formed by

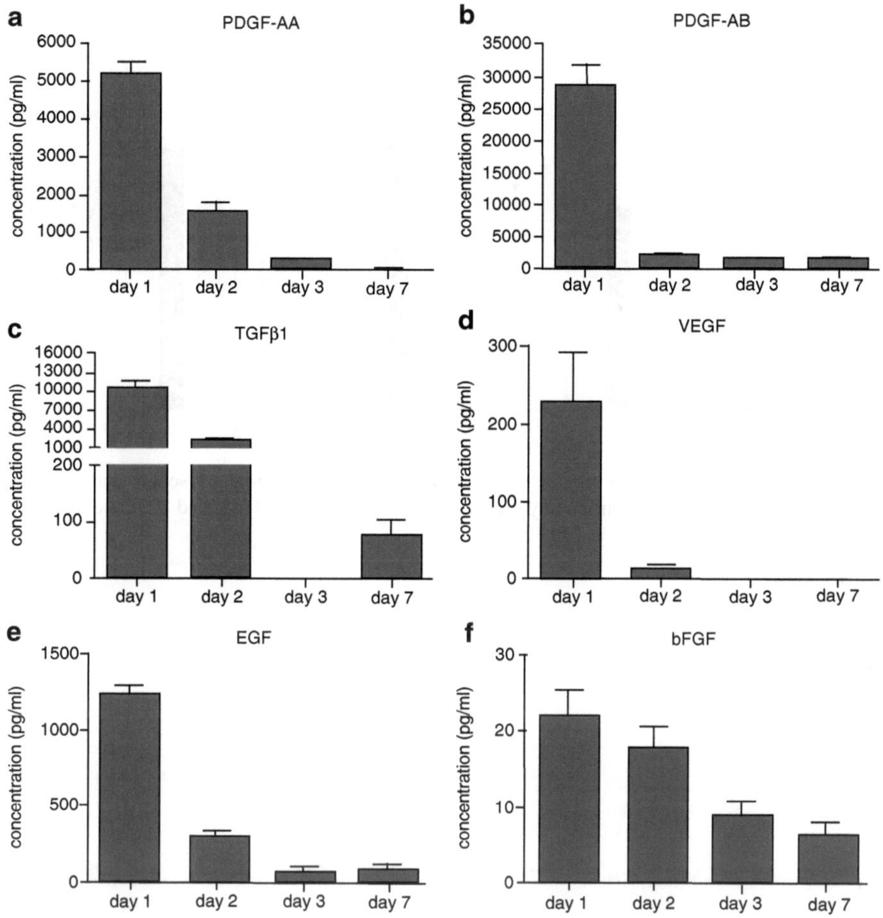

Fig. 7.29 Quantification of growth factor release in PRFM-conditioned media (CM) obtained from three donors. (**a**) PDGF-AA, (**b**) PDGF-AB, (**c**) TGF-β1, (**d**) VEGF, (**e**) EGF, and (**f**) bFGF [67]

adding PRP to alginate solution, and subsequently dropping into calcium chloride solution, while capsules were created by mixing PRP with calcium chloride and then dropping into the alginate solution. In vitro release profiles measured for 7 days show that significantly higher concentrations of TGF-β1 were released from the beads than from the capsules, although the release of PDGF and IGF-1 was significantly higher from the capsules, as demonstrated in Fig. 7.32. The growth of osteoblast-like cells was significantly higher at day 3 when cultured with alginate beads, and at day 14 in the capsule group (Fig. 7.33). TGF-β1 is thought to be the responsible mitogenic factor as its release profile follows a similar trend. The ALP activity at day 1 was higher for the capsule group due to the higher IFG-1 release from this system. The comparison of TGF-β1 release from different PRP carriers is shown in Fig. 7.34. PRP activated with thrombin released 90% of TGF-β1 within 2 h. The addition of bone substitutes, such as allograft (AG) and BioOss, to PRP

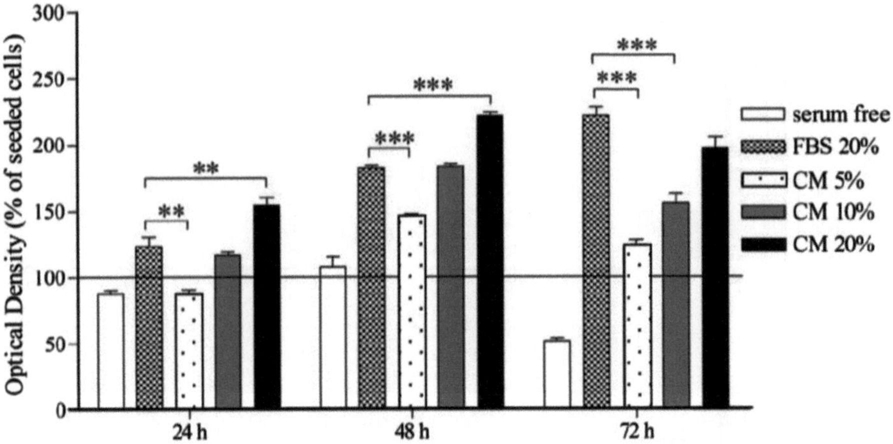

Fig. 7.30 PRFM-conditioned media (CM)-induced cell proliferation. Mesenchymal stem cell (MSC) proliferation was determined by methylene blue assay after 24, 48, and 72 h in culture [67]

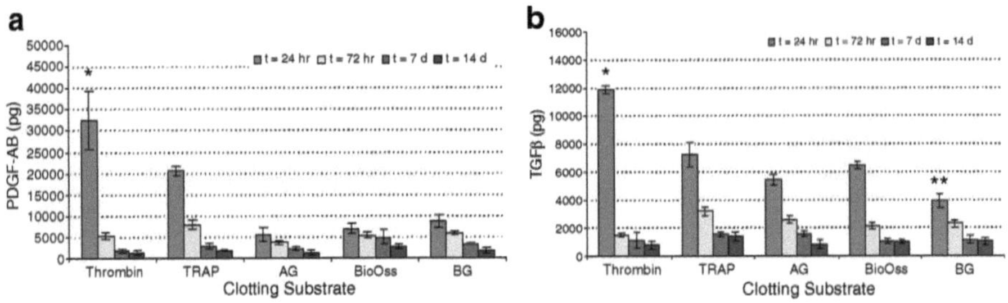

Fig. 7.31 Effects of clotting substrate on release of (**a**) PDGF and (**b**) TGF-β [68]

reduced TGF-β1 release by 50%. From the alginate carriers, less than 10% of factors were released in 2 h. It was concluded that the release of PRP-derived growth factors could be tailored by altering the carrier type, in order to control cellular responses.

Gelatin microspheres have been clinically used as carrier for sustained release of growth factors. Bir et al. investigated the effect of PRP solution (PRP-sol) versus the sustained release of growth factors from gelatin hydrogel (PRP-sr), on perfusion and neovascularization in a murine model of hind-limb ischemia [72]. PRP was prepared by double centrifugation and was absorbed into freeze-dried gelatin microspheres by soaking for 1 h at 37°C. It is likely that platelets are activated during impregnation of the gelatin molecules, with the secreted growth factors being immobilized in the hydrogel through a physicochemical interaction. Growth factors may be released from the hydrogel as a result of hydrogel degradation. Figure 7.35

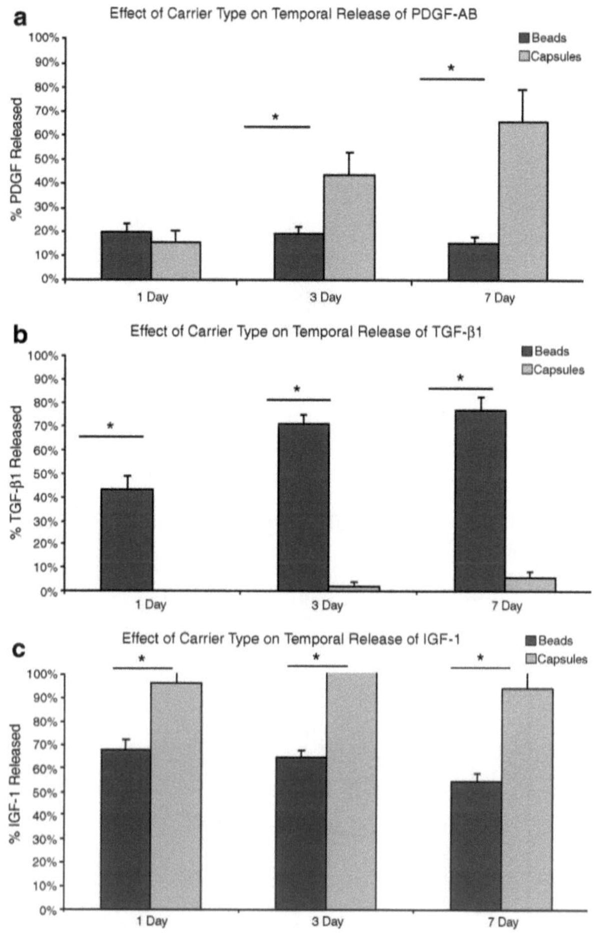

Fig. 7.32 Effects of carrier type on the release of PRP-derived growth factors [71]

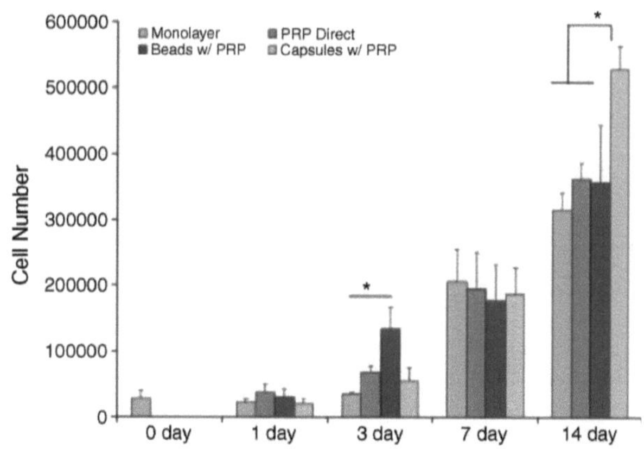

Fig. 7.33 Bioactivity of factors released from alginate carriers. The PRP-derived growth factors released from the carriers remained bioactive and promoted carrier-dependent increases in the proliferation of osteoblast-like cells [71]

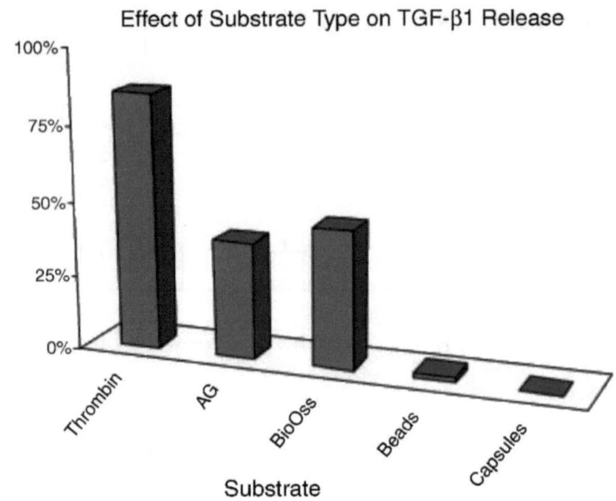

Fig. 7.34 Release of TGF-β1 from alginate carriers compared to current methods of PRP preparation [71]

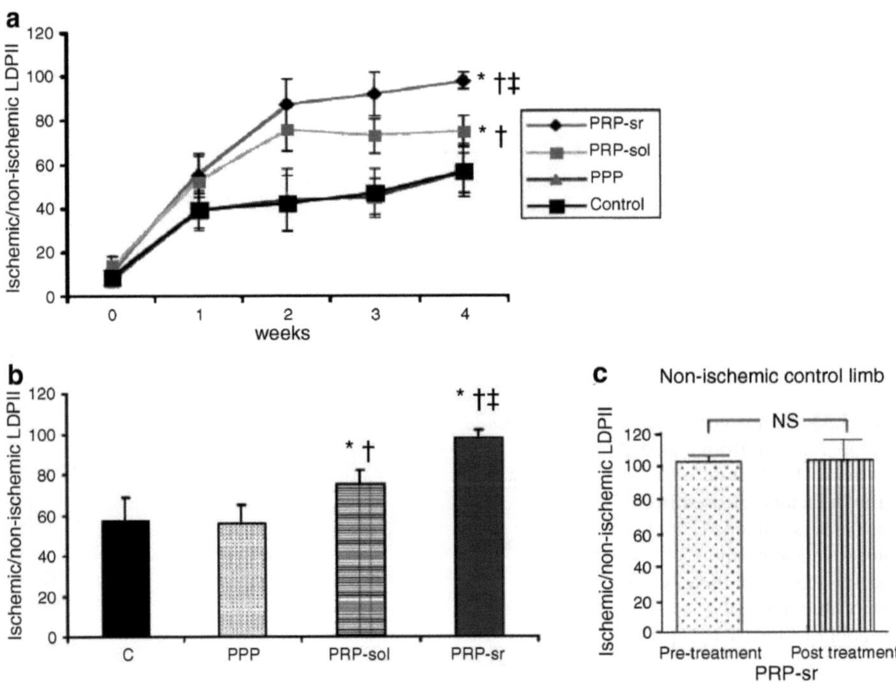

Fig. 7.35 (**a**) Time course of ischemic/nonischemic blood perfusion ratio in different groups of mice subjected to various treatments. (**b**) Ischemic/nonischemic blood perfusion ratio in the different groups at 4 weeks after various treatments. (**c**) Blood perfusion ratio in contralateral nonischemic hind limb before and 4 weeks after treatment with PRP-sr [72]

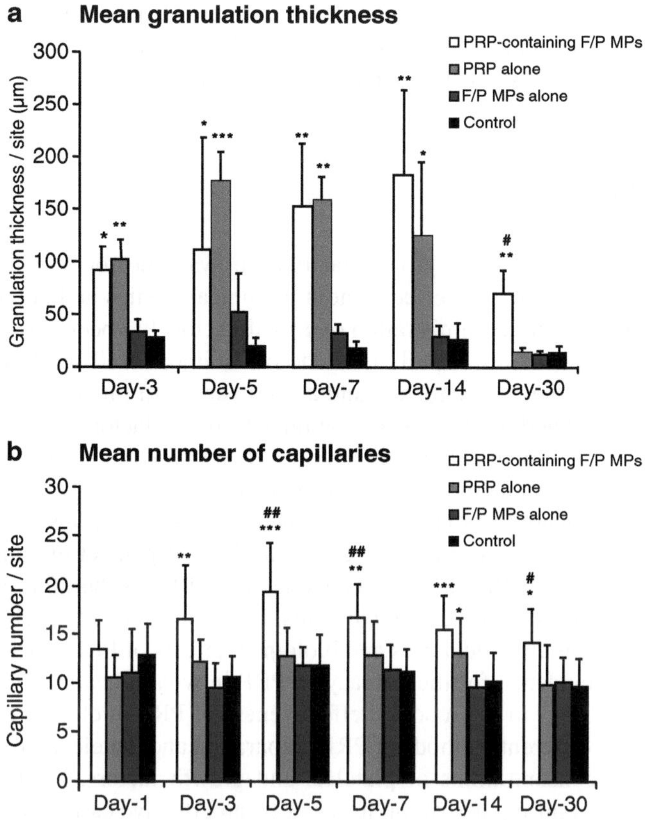

Fig. 7.36 Changes in granulation thickness (**a**) and number of capillaries (**b**) at sites injected with PRP-containing fragmin/protamine microparticles [73]

represents the time course of ischemic/nonischemic blood perfusion ratio (a), ischemic/nonischemic blood perfusion ratio at 4 weeks (b), and blood perfusion ratio in contralateral nonischemic hind limb before, and 4 weeks after, treatment with PRP-sr (c). As is apparent from Fig. 7.35, the sustained release of potent angiogenic growth factors contained in PRP was found to restore blood perfusion, perhaps by stimulating angiogenesis, arteriogenesis, and vasculogenesis in mouse hind-limb ischemia.

The effect of fragmin/protamine microparticles containing PRP as a protein carrier was evaluated in neovascularization and granulation tissue formation [73]. PRP was prepared by double centrifugation and then frozen at −80°C. The thawed PRP was mixed with the microparticles before subcutaneous injection into the backs of mice. Changes in granulation thickness and number of capillaries at the implantation site were evaluated from day 3 to day 30. Figure 7.36 indicates that both PRP-containing microparticles and PRP alone induced significant formation of granulation tissue at the site, while only the PRP-containing microparticle group maintained the newly formed granulation tissue for 30 days, demonstrating the efficacy of the microparticles as a slow-release PRP delivery system.

Conclusion

Although many recombinant growth factors are available for research purposes, only a few have shown their effectiveness in clinical studies and are available for human therapies. Bolus administration of growth factors has been shown to be ineffective as they diffuse away rapidly from the site of injection. Therefore, the importance of DDSs has been recognized, and a wide variety of bioabsorbable materials have been investigated as carriers. Recombinant growth factors are expensive, and the use of a single growth factor may not be sufficient for wound-healing processes where multiple growth factors must be responsible.

PRP can be prepared from the patient's own plasma, and hence, it is completely safe and relatively inexpensive. High concentrations of platelet growth factors are released from PRP to enhance wound healing, and PRP has therefore been applied to a variety of tissues in broad surgical fields. However, more research is needed to establish the potential clinical uses and biological effects of PRP, since there is still limited clinical evidence for the efficacy of PRP in wound healing. There are also some inconsistencies in the reported effectiveness of PRP in clinical trials. This is probably due to different methods of PRP preparation and platelet activation, resulting in various concentrations of platelets and growth factors. Further, the weak fibrin clot formed during activation may not be an effective carrier for growth factors. PRF with a strong fibrin clot and use of bioabsorbable materials as carriers may be required for effective clinical applications.

References

1. Urist MR (1965) Bone: formation by autoinduction. Science 150(698):893–9
2. White AP, Vaccaro AR, Hall JA, Whang PG, Friel BC, McKee MD (2007) Clinical applications of BMP-7/OP-1 in fractures, non-unions and spinal fusion. Int Orthop 31:735–41
3. McKay WF, Peckham SM, Badura JM (2007) A comprehensive clinical review of recombinant human bone morphogenetic protein-2 (INFUSE® bone graft). Int Orthop 31:729–34
4. Tabata Y, Yamamoto M, Ikada Y (1998) Biodegradable hydrogels for bone regeneration through growth factor release. Pure Appl Chem 70(6):1277–82
5. Coelho JF, Ferreira PC, Alves P, Cordeiro R, Fonseca AC, Góis JR, Gil MH (2010) Drug delivery systems: advanced technologies potentially applicable in personalized treatments. EPMA J 1:164–209
6. Bajpai AK, Shukla SK, Bhanu S, Kankane S (2008) Responsive polymers in controlled drug delivery. Prog Polym Sci 33:1088–118
7. Anitua E, Sánchez M, Orive G, Andia I (2008) Delivering growth factors for therapeutics. Trends Pharmacol Sci 29(1):37–41

8. Langer R, Tirrell DA (2004) Designing materials for biology and medicine. Nature 428(6982):487–92
9. Li RH, Wozney JM (2001) Delivering on the promise of bone morphogenic proteins. Trends Biotechnol 19(7):255–65
10. Burkus JK, Heim SE, Gornet MF, Zdeblick TA (2003) Is INFUSE® bone graft superior to autograft bone? An integrated analysis of clinical trials using the LT-CAGE lumbar tapered fusion device. J Spinal Disord Tech 16(2)):113–22
11. Jones AL, Bucholz RW, Bosse MJ, Mirza SK, Lyon TR, Webb LX, Pollak AN, Golden JD, Valentin-Opran A, BMP-2 Evaluation in Surgery for Tibial Trauma-Allgraft (BESTT-ALL) Study Group (2006) Recombinant human BMP-2 and allograft compared with autogenous bone graft for reconstruction of diaphyseal tibial fractures with cortical defects A randomized, controlled trial. J Bone Joint Surg Am 88:1431–41
12. OP-1 Putty (2011) [updated 2011; cited 2011 08 June]; Available from: http://www.op1.com/product_op-1_putty.html
13. Friedlaender GE, Perry CR, Cole JD, Cook SD, Cierny G, Muschler GF, Zych GA, Calhoun JH, LaForte AJ, Yin S (2001) Osteogenic protein-1 (Bone morphogenetic protein-7) in the treatment of tibial nonunions. J Bone Joint Surg Am 83(Suppl 1 (Pt2)):S151–S8
14. Vaccaro AR, Patel T, Fischgrund J, Anderson DG, Truumees E, Herkowitz HN, Phillips F, Hilibrand A, Albert TJ, Wetzel T, McCulloch JA (2004) A pilot study evaluating the safety and efficacy of OP-1 Putty (rhBMP-7) as a replacement for iliac crest autograft in posterolateral lumbar arthrodesis for degenerative spondylolisthesis. Spine 29(17):1885–92
15. Vaccaro AR, Anderson DG, Patel T, Fischgrund J, Truumees E, Herkowitz HN, Phillips F, Hilibrand A, Albert TJ, Wetzel T, McCulloch JA (2005) Comparison of OP-1 Putty (rhBMP-7) to iliac crest autograft for posterolateral lumbar arthrodesis: a minimum 2-year follow-up pilot study. Spine (Phila Pa 1976) 30(24):2709–16
16. Lee YM, Nam SH, Seol YJ, Kim TI, Lee SJ, Ku Y, Rhyu IC, Chung CP, Han SB, Choi SM (2003) Enhanced bone augmentation by controlled release of recombinant human bone morphogenetic protein-2 from bioabsorbable membranes. J Periodontol 74(6):865–72
17. Ikada Y (2006) Challenges in tissue engineering. J R Soc Interface 3:589–601
18. Yamamoto M, Takahashi Y, Tabata Y (2003) Controlled release by biodegradable hydrogels enhances the ectopic bone formation of bone morphogenetic protein. Biomaterials 24(24): 4375–83
19. Okamoto T, Yamamoto Y, Gotoh M, Huang CL, Nakamura T, Shimizu Y, Tabata Y, Yokomise H (2004) Slow release of bone morphogenetic protein 2 from a gelatin sponge to promote regeneration of tracheal cartilage in a canine model. J Thorac Cardiovasc Surg 127:329–34
20. Asamura S, Mochizuki Y, Yamamoto M, Tabata Y, Isogai N (2010) Bone regeneration using a bone morphogenetic protein-2 saturated slow-release gelatin hydrogel sheet: evaluation in a canine orbital floor fracture model. Ann Plast Surg 64(4):496–502
21. Boerckel JD, Kolambkar YM, Dupont KM, Uhrig BA, Phelps EA, Stevens HY, García AJ, Guldberg RE (2011) Effects of protein dose and delivery system on BMP-mediated bone regeneration. Biomaterials 32(22):5241–51
22. Hankemeier S, Keus M, Zeichen J, Jagodzinski M, Barkhausen T, Bosch U, Krettek C, Van Griensven M (2005) Modulation of proliferation and differentiation of human bone marrow stromal cells by fibroblast growth factor 2: potential implications for tissue engineering of tendons and ligaments. Tissue Eng 11(1–2):41–9
23. Nakamura S, Kanatani Y, Kishimoto S, Nakamura S, Ohno C, Horio T, Masanori F, Hattori H, Tanaka Y, Kiyosawa T, Maehara T, Ishihara M (2009) Controlled release of FGF-2 using fragmin/protamine microparticles and effect on neovascularization. J Biomed Mater Res A 91(3):814–23
24. Phelps EA, Landázuri N, Thulé PM, Taylor WR, García AJ (2010) Bioartificial matrices for therapeutic vascularization. Proc Natl Acad Sci USA 107:3323–8
25. Miyagi Y, Chiu LL, Cimini M, Weisel RD, Radisic M, Li RK (2011) Biodegradable collagen patch with covalently immobilized VEGF for myocardial repair. Biomaterials 32(5):1280–90

26. Richardson TP, Peters MC, Ennett AB, Mooney DJ (2001) Polymeric system for dual growth factor delivery. Nat Biotech 19:1029–34
27. Branco MC, Pochan DJ, Wagner NJ, Schneider JP (2010) The effect of protein structure on their controlled release from an injectable peptide hydrogel. Biomaterials 31(36):9527–34
28. Shen H, Hu X, Yang F, Bei J, Wang S (2011) Cell affinity for bFGF immobilized heparin-containing poly(lactide-co-glycolide) scaffolds. Biomaterials 32(13):3404–12
29. Marx RE (2004) Platelet-rich plasma: evidence to support its use. J Oral Maxillofac Surg 62:489–96
30. Dohan Ehrenfest DM, Rasmusson L, Albrektsson T (2009) Classification of platelet concentrates: from pure platelet-rich plasma (P-PRP) to leucocyte and platelet-rich fibrin (L-PRF). Trends Biotechnol 27(3):158–67
31. Willoughby S, Holmesa A, Loscalzo J (2002) Review: platelets and cardiovascular disease. Eur J Cardiovasc Nurs 1(4):273–88
32. Dohan DM, Choukroun J, Diss A, Dohan SL, Dohan AJJ, Mouhyi J, Gogly B (2006) Platelet-rich fibrin (PRF): a second-generation platelet concentrate. Part I: technological concepts and evolution. Oral Surg Oral Med Oral Pathol Oral Radiol Endo 101:E37–44
33. Eppley BL, Woodell JE, Higgins J (2004) Platelet quantification and growth factor analysis from platelet-rich plasma: implications for wound healing. PRS 114(6):1502–8
34. Chouhan VD, De La Cadena RA, Nagaswami C, Weisel JW, Kagani M, Rao AK (1997) Simultaneous occurrence of human antibodies directed against fibrinogen, thrombin, and factor V following exposure to bovine thrombin: effects on blood coagulation, protein C activation and platelet function. Thrombisus Haenistasis 77(2):343–9
35. Marx RE (2001) Platelet-rich plasma (PRP): what is PRP and what is not PRP? Implant Dent 10(4):225–8
36. Fufa D, Shealy B, Jacobson M, Kevy S, Murray MM (2008) Activation of platelet-rich plasma using soluble type I collagen. J Oral Maxillofac Surg 66(4):684–90
37. Thorn JJ, Sørensen H, Weis-Fogh U, Andersen M (2004) Autologous fibrin glue with growth factors in reconstructive maxillofacial surgery. Int J Oral Maxillofac Surg 33(1):95–100
38. Lacoste E, Martineau I, Gagnon G (2003) Platelet concentrates: effects of calcium and thrombin on endothelial cell proliferation and growth factor release. J Periodontol 74(10):1498–507
39. Weibrich G, Kleis WKG, Hafner G, Hitzler WE (2002) Growth factor levels in platelet-rich plasma and correlations with donor age, sex, and platelet count. J Cranio Maxillofac Surg 30(2):97–102
40. Weibrich G, Kleis WK, Hafner G (2002) Growth factor levels in the platelet-rich plasma produced by 2 different methods: curasan-type PRP kit versus PCCS PRP system. Int J Oral Maxillofac Implants 17(2):184–90
41. Harrison S, Vavken P, Kevy S, Jacobson M, Zurakowski D, Murray MM (2011) Platelet activation by collagen provides sustained release of anabolic cytokines. Am J Sports Med 39(4):729–34
42. Kakudo N, Minakata T, Mitsui T, Kushida S, Notodihardjo FZ, Kusumoto K (2008) Proliferation-promoting effect of platelet-rich plasma on human adipose-derived stem cells and human dermal fibroblasts. Plast Reconstr Surg 122(5):1352–60
43. Graziani F, Ivanovski S, Cei S, Ducci F, Tonetti M, Gabriele M (2006) The in vitro effect of different PRP concentrations on osteoblasts and fibroblasts. Clin Oral Implants Res 17(2):212–9
44. Choi BH, Zhu SJ, Kim BY, Huh JY, Lee SH, Jung JH (2005) Effect of platelet-rich plasma (PRP) concentration on the viability and proliferation of alveolar bone cells: an in vitro study. Int J Oral Maxillofac Surg 34(4):420–4
45. Liu Y, Kalén A, Risto O, Wahlström O (2002) Fibroblast proliferation due to exposure to a platelet concentrate in vitro is pH dependent. Wound Repair Regen 10(5):336–40
46. Anitua E, Andia I, Ardanza B, Nurden P, Nurden AT (2004) Autologous platelets as a source of proteins for healing and tissue regeneration. Thromb Haemost 91(1):4–15
47. Weibrich G, Hansen T, Kleis W, Buch R, Hitzler WE (2004) Effect of platelet concentration in plattelet-rich plasma on peri-implant bone regeneration. Bone 34:665–71
48. Phillips GD, Stone AM, Whitehead RA, Knighton DR (1994) Platelet derived wound healing factors (PDWHF) accelerate and augment wound healing angiogenesis in the rat. In Vivo 8(2):167–71

49. Cho JM, Lee YH, Baek RM, Lee SW (2011) Effect of platelet-rich plasma on ultraviolet b-induced skin wrinkles in nude mice. J Plast Reconstr Aesthet Surg 64(2):e31–9
50. Oh DS, Cheon YW, Jeon YR, Lew DH (2011) Activated platelet-rich plasma improves fat graft survival in nude mice: a pilot study. Dermatol Surg 37(5):619–25
51. Zechner W, Tangl S, Tepper G, Fürst G, Bernhart T, Haas R, Mailath G, Watzek G (2003) Influence of platelet-rich plasma on osseous healing of dental implants: a histologic and histomorphometric study in minipigs. Int J Oral Maxillofac Implants 18(1):15–22
52. Kim SG, Chung CH, Kim YK, Park JC, Lim SC (2002) Use of particulate dentin-plaster of Paris combination with/without platelet-rich plasma in the treatment of bone defects around implants. Int J Oral Maxillofac Implants 17:86–94
53. Marx RE, Carlson ER, Eichstaedt RM, Schimmele SR, Strauss JE, Georgeff KR (1998) Platelet-rich plasma: growth factor enhancement for bone grafts. Oral Surg Oral Med Oral Pathol Oral Radiol Endo 85:638–46
54. Cervelli V, Gentile P, Scioli MG, Grimaldi M, Casciani CU, Spagnoli LG, Orlandi A (2009) Application of platelet-rich plasma in plastic surgery: clinical and in vitro evaluation. Tissue Eng C 15:1–10
55. Saad Setta H, Elshahat A, Elsherbiny K, Massoud K, Safe I (2011) Platelet-rich plasma versus platelet-poor plasma in the management of chronic diabetic foot ulcers: a comparative study. Int Wound J 8(3):307–12
56. Chen TM, Tsai JC, Burnouf T (2010) A novel technique combining platelet gel, skin graft, and fibrin glue for healing recalcitrant lower extremity ulcers. Dermatol Surg 36(4):453–60
57. Messora MR, Nagata MJ, Dornelles RC, Bomfim SR, Furlaneto FA, de Melo LG, Deliberador TM, Bosco AF, Garcia VG, Fucini SE (2008) Bone healing in critical-size defects treated with platelet-rich plasma activated by two different methods. A histologic and histometric study in rat calvaria. J Periodontal Res 43(6):723–9
58. Rodriguez A, Anastassov GE, Lee H, Buchbinder D, Wettan H (2003) Maxillary sinus augmentation with deproteinated bovine bone and platelet rich plasma with simultaneous insertion of endosseous implants. J Oral Maxillofac Surg 61(2):157–63
59. Choi BH, Im CJ, Huh JY, Suh JJ, Lee SH (2004) Effect of platelet-rich plasma on bone regeneration in autogenous bone graft. Int J Oral Maxillofac Surg 33(1):56–9
60. Man D, Plosker H, Winland-Brown JE (2001) The use of autologous platelet-rich plasma (platelet gel) and autologous platelet-poor plasma (fibrin glue) in cosmetic surgery. Plast Reconstr Surg 107(1):229–37
61. Thor A, Franke-Stenport V, Johansson CB, Rasmusson L (2007) Early bone formation in human bone grafts treated with platelet-rich plasma: preliminary histomorphometric results. Int J Oral Maxillofac Surg 36(12):1164–71
62. Mazor Z, Horowitz RA, Del Corso M, Prasad HS, Rohrer MD, Dohan Ehrenfest DM (2009) Sinus floor augmentation with simultaneous implant placement using Choukroun's platelet-rich fibrin as the sole grafting material: a radiologic and histologic study at 6 months. J Periodontol 80(12):2056–64
63. Dohan Ehrenfest DM, de Peppo GM, Doglioli P, Sammartino G (2009) Slow release of growth factors and thrombospondin-1 in Choukroun's platelet-rich fibrin (PRF): a gold standard to achieve for all surgical platelet concentrates technologies. Growth Factors 27(1):63–9
64. Kang YH, Jeon SH, Park JY, Chung JH, Choung YH, Choung HW, Kim ES, Choung PH (2011) Platelet-rich fibrin is a Bioscaffold and reservoir of growth factors for tissue regeneration. Tissue Eng Part A 17(3–4):349–59
65. Huang FM, Yang SF, Zhao JH, Cgang YC (2010) Platelet-rich fibrin increases proliferation and differentiation of human dental pulp cells. J Endod 36(10):1628–32
66. Gassling V, Douglas T, Warnke PH, Açil Y, Wiltfang J, Becker ST (2010) Platelet-rich fibrin membranes as scaffolds for periosteal tissue engineering. Clin Oral Implants Res 21(5):543–9
67. Lucarelli E, Beretta R, Dozza B, Tazzari PL, O'Connel SM, Ricci F, Pierini M, Squarzoni S, Pagliaro PP, Oprita EI, Donati D (2010) A recently developed bifacial platelet-rich fibrin matrix. Eur Cell Mate 20:13–23

68. Tsay RC, Vo J, Burke A, Eisig SB, Lu HH, Landesberg R (2005) Differential growth factor retention by platelet-rich plasma composites. J Oral Maxillofac Surg 63:521–8
69. Coughlin SR (1999) Protease-activated receptors and platelet function. Thromb Haemost 82(2):353–6
70. Stiernberg J, Redin WR, Warner WS, Carney DH (1993) The role of thrombin and thrombin receptor activating peptide (TRAP-508) in initiation of tissue repair. Thromb Haemost 70(1): 158–62
71. Lu HH, Vo JM, Chin HS, Lin J, Cozin M, Tsay R, Eisig S, Landesberg R (2008) Controlled delivery of platelet-rich plasma-derived growth factors for bone formation. J Biomed Mater Res A 86(4)):1128–36
72. Bir SC, Esaki J, Marui A, Yamahara K, Tsubota H, Ikeda T, Sakata R (2009) Angiogenic properties of sustained release platelet-rich plasma: characterization in-vitro and in the ischemic hind limb of the mouse. J Vasc Surg 50(4):870–9
73. Takikawa M, Nakamura S, Nakamura S, Nambu M, Ishihara M, Fujita M, Kishimoto S, Doumoto T, Yanagibayashi S, Azuma R, Yamamoto N, Kiyosawa T (2011) Enhancement of vascularization and granulation tissue formation by growth factors in human platelet-rich plasma-containing fragmin/protamine microparticles. J Biomed Mater Res B Appl Biomater 97(2):373–80

Chapter 8
Sutures for Wound Closure

Abstract For over a century, sutures have been almost exclusively used for wound closure and remain the largest group of biomaterials used for surgical operations. Since the first introduction of synthetic, bioabsorbable polymers in the 1970s, they have found successful applications as suturing materials. Today, a wide variety of sutures made of both nonbioabsorbable and bioabsorbable polymers are clinically available. The requirements for sutures depend on the tissues to which they are applied, and therefore, surgeons need to choose appropriate sutures depending on the surgical procedure. Bioabsorbable polymers used for suture materials include poly(glycolic acid) (PGA), poly-*p*-dioxanone, and their copolymers alone. This chapter focuses on the characteristics of commercially available bioabsorbable sutures and their applications after classification of the thread materials.

Introduction

After an injury or surgery, a surgical suture is used to hold tissues together. A suture consists of a needle with a length of thread attached. The optimal suture should be easy to handle and have high tensile strength and knot security. It should cause minimal tissue reaction, and its material should resist infection and have good elasticity and plasticity in order to accommodate wound swelling. However, there is no single suture that can fulfill all these criteria. Therefore, a surgeon must choose suture material based on type of surgery that she or he is performing because different tissues have different requirements for suture support (some need only a few days, e.g., muscle, subcutaneous tissue, and skin, while others require weeks or even months, e.g., fascia and tendons). In addition, the healing rates of tissues differ depending on factors such as infection, debility, respiratory problems, obesity, collagen disorders, malnutrition, malignancy, and drugs. Often, the trade-off is tissue handling versus longevity versus healing properties. General classifications of thread material are natural and synthetic, bioabsorbable and nonbioabsorbable, and

S. Suzuki and Y. Ikada, *Biomaterials for Surgical Operation*,
DOI 10.1007/978-1-61779-570-1_8,

Table 8.1 Suture characteristics [1]

Characteristic	Definition
Coefficient of friction	A suture's relative resistance to being passed through a tissue. A higher coefficient of friction leads to increased local tissue damage. Monofilaments usually have the lowest coefficient of friction
Tensile strength	A suture's ability to resist breakage. The strength of the suture should not be significantly greater than the tissue it is approximating (if the tensile strength is too high, the suture will cut through tissue)
Elasticity	A suture's ability to regain its original length after deformation. Elasticity should be sufficient to accommodate tissue swelling
Memory	A suture's ability to return to its original shape after deformation. Similar to pliability (more memory = less pliability and less knot security)
Cost	A significant issue in today's health-care arena. Newer suture materials with precision needles are generally more costly

monofilament and multifilament. Mechanical properties can be further divided in terms of the multiple characteristics of each material, as shown in Table 8.1 [1].

Originally, sutures were made from biological materials such as catgut and silk. J. Lister first attempted sterilization in the 1860s with "carbolic catgut," and the chromic catgut followed two decades later. Sterile catgut was finally achieved in 1906 with iodine treatment. This became the standard method of suture preparation for nearly half a century. The first synthetic bioabsorbable suture was made from PVA in 1931. Polyesters were developed in the 1950s, and later, the process of radiation sterilization was established for catgut and polyester. Today, most sutures are made of synthetic polymer fibers. Silk and gut sutures are the only materials still—though rarely—in use from ancient times. In fact, gut sutures have been banned in Europe and Japan because of concerns of bovine spongiform encephalopathy.

Nonbioabsorbable Sutures

Nonbioabsorbable sutures are defined by their resistance to degradation by living tissues. They are most useful in percutaneous closures. Synthetic, nonbioabsorbable, monofilament sutures include nylon, polypropylene, and polybutester sutures, while synthetic, nonbioabsorbable, multifilament (braided) sutures are composed of nylon and polyester. Polybutester, developed in 2000, is a block copolymer that contains butylene terephthalate and tetramethylene ether glycol. Metallic fibers such as steel fibers are also used extensively for suturing.

Bioabsorbable Sutures

Bioabsorbable sutures are defined by the loss of most of their strength within 60 days after placement. They provide temporary wound support until the wound heals well enough to withstand normal stress and are used primarily as buried sutures to close

Table 8.2 Commercially available synthetic absorbable sutures

	Chemical structure	Manufacturer	Trade name	Introduced year	Complete hydrolysis
Multifilament (Braided)	Polyglycolide	COVIDIEN	DEXON II	1970	90–120 days[a]
	Poly(glycolide-*co*-L-lactide) (90:10)	Johnson and Johnson	VICRYL	1974	56–70 days[b]
	Poly(glycolide-*co*-L-lactide) (90:10)	COVIDIEN	Polysorb	1981	
Monofilament	Poly-*p*-dioxanone	Johnson and Johnson	PDS II	1982	180–210 days
	Poly(glycolide-*co*-trimethylene carbonate)	COVIDIEN	Maxon	1985	180–210 days
	Poly(glycolide-*co*-ε-caprolactone)	Johnson and Johnson	Monocryl	1993	91–119 days
	Poly(glycolide-*co*-trimethylene carbonate-*co*-*p*-dioxanone)	COVIDIEN	Biosyn	1999	90–110 days
	Poly(glycolide-*co*-trimethylene carbonate)	B.Braun	Monosyn	2000	
	Poly(glycolide-*co*-ε-caprolactone-*co*-trimethylene carbonate-*co*-L-lactide)	COVIDIEN	Caprosyn	2002	56 days

[a]42 days for a fast-absorbing polyglycolide suture (Polysyn FA)
[b]35 days for γ-irradiated poly(glycolide-*co*-L-lactide) suture (Vicryl Rapid 1987)

the dermis and subcutaneous tissue and reduce wound tension. When the Young's modulus of the thread materials is relatively low, they are processed into monofilament sutures, while multifilament sutures are made of less pliable materials and processed into twisted or braided construction.

Multifilament sutures generally have greater tensile strength and better pliability and flexibility than monofilaments. Multifilament sutures with a high coefficient of friction are more difficult to pass through tissue and cause a greater degree of tissue injury during placement and removal. However, these sutures are easier to handle and manipulate for tying knots. The knots of monofilament sutures are smaller than those of multifilaments. Multifilament sutures have a high degree of capillarity, which is correlated with a tendency to absorb and retain both fluid and bacteria. This may promote infection if bacterial contamination occurs during or shortly after surgery.

Synthetic, bioabsorbable sutures available today are listed in Table 8.2, together with their chemical structure. As can be seen, they are composed of copolymers, except for PGA and poly-*p*-dioxanone. One of the monomers in the copolymers of all sutures is glycolide. The surface of most multifilament sutures is coated to permit easy tissue passage, precise knot placement, and smooth tie-down. The coating materials applied include calcium stearate, poly(ε-caprolactone) (PCL), PGLA (30:70), and poly(CL-co-GA). The characteristics of these bioabsorbable sutures are briefly described below.

PGA

This first synthesized bioabsorbable suture has high tensile strength, with a retention of 60% at day 7, 35% at day 14, and only 5% at day 28. Uncoated PGA sutures have good handling and knot security properties, but their high coefficient of friction results in significant tissue drag. Coating with PCL minimizes this drag, but four throws are recommended to ensure secure knots.

PGLA

PGLA is synthesized by means of random ring-opening polymerization of GA and LLA. Depending on the ratio of LLA to GA used for the polymerization, different forms of PGLA can be obtained. Multifilament braided Vicryl® sutures contain 90/10 molar ratio of GA to LLA, and they are coated with 2–10% of a 50:50 mixture of an amorphous PGLA copolymer of 35/65 mole ratio and calcium stearate. The initial tensile strength of the copolymer suture is slightly greater than that of PGA. The PGLA suture retains 60% of its tensile strength at day 14 after implantation and only 8% of its original strength at day 28. Tissue reactivity with PGLA is low, less than that of PGA. The water-repelling quality of the lactide unit may slow the loss of tensile strength, and the bulkiness of the lactide unit leads to rapid bioabsorption of the suture mass once tensile strength is lost. The PGLA sutures are used in general soft tissue approximation and vessel ligation without adverse outcomes and resultant cost savings.

Poly-p-dioxanone

Multifilament sutures such as PGA and PGLA develop a greater amount of friction when penetrating tissues and have a higher risk of infection. Consequently, monofilament sutures made of polydioxanone that had a smooth and soft surface were introduced in the 1980s. Polydioxanone is a colorless, semicrystalline polymer with a very low glass transition temperature, ranging from −10°C to 0°C. In the body, this polymer is broken down into glycoxylate and excreted in the urine or converted into glycine and subsequently into carbon dioxide and water. Although the initial tensile strength of the polydioxanone suture is lower than that of the two synthetic multifilament sutures mentioned above, it retains its strength longer. At day 14 after implantation, it has 74% residual initial strength, 58% at day 28, and 41% at week 6. The initial polydioxanone suture was stiff and had poor handling and knot-tying properties, but the newer product (PDS-II) has improved handling capabilities and has replaced the original product. The polydioxanone suture is useful as a buried suture in wounds that require prolonged dermal support and in contaminated wounds or wounds in locations at greater risk for infection. This suture provides extended wound support and elicits only a slight tissue reaction. Polydioxanone is more expensive than PGA and PGLA sutures.

Poly(GA-co-TMC)

This type of copolymer is prepared as A–B–A block copolymers in a 2:1 GA:TMC ratio, with a GA–TMC center block (B) and pure GA end blocks (A). These materials have better flexibility than pure PGA and are absorbed in approximately 7 months. This copolymer was developed to combine the predictable in vivo performance of a synthetic absorbable suture with the handling characteristics of a monofilament suture. The copolymer has a high initial tensile strength (greater than that of polydioxanone) and retains 81% of its strength at day 14, 59% at day 28, and 30% at week 6. This suture is easier to handle and has greater knot security than the three bioabsorbable sutures mentioned above.

Poly(GA-co-CL)

The biodegradability of this copolymer is much higher than that of the polydioxanone suture. The higher rate of degradation of poly(GA-co-CL) may be due to its GA content. This copolymer suture has superior pliability despite its monofilament nature, leading to easy handling and tying. Among all bioabsorbable monofilament sutures, this has the highest tensile strength, which is high initially, 50–60% at day 7, but only 20–30% at day 14 after implantation. This suture is used for subcuticular closure and is most useful as a buried suture in wounds that do not require prolonged dermal support.

Poly(GA-co-TMC-co-p-dioxanone)

This copolymer suture has greater tensile strength than the PGLA suture 4 weeks after implantation. The handling characteristics and knot security are also superior. Tissue drag and risk of bacterial infection are lower than those of PGLA, similar to the case with monofilament construction. This suture retains 75% of its initial strength at day 14 and 40% at day 21.

Poly(GA-co-CL-co-TMC-co-LLA)

This newest synthetic bioabsorbable suture is a rapidly degradable monofilament polyester. When compared with the chromic gut suture, this synthetic suture has high tensile strength, low tissue reactivity, and improved handling characteristics. Its greatest advantage is its rapid rate of bioabsorption. Animal studies indicate that this suture retains a minimum of 50–60% knot strength at day 5 and a minimum of 20–30% knot strength at day 10 after implantation. It provides secure wound approximation for 10 days, and all tensile strength is lost by day 21.

Others

Wound infection is considered to be one of the oldest and most common complications in all types of injuries. The presence of foreign materials in a wound has been known to increase the susceptibility of surrounding tissues to wound infection. Suture materials are probably the most important biomaterials in wound infection because the infection begins along or near suture lines. Incorporation of antibacterials is used to impart antimicrobial activity to biomaterials. An antibacterial form is also available for PGLA and poly(GA-co-CL) sutures. The antibacterial agent used to coat sutures is triclosan (2,4,4′-trichroro-2′-hydroxydiphenyl ether). Figure 8.1 shows the structural formula. This active substance, which is slightly soluble in water, was introduced in 1965. Triclosan inhibits fatty acid biosynthesis and enoylacyl carrier protein reductase, resulting in membrane destabilization. This antibacterial agent has been shown to inhibit colonization on the suture by methicillin-sensitive and methicillin-resistant *Staphylococcus aureus* and *S. epidermis*, *Escherichia coli*, and *Klebsiella pneumoniae*, even after direct in vivo challenge with bacteria. The handling and wound healing characteristics and bioabsorption profile after coating with this nontoxic and nonirritating agent are similar to those with untreated sutures.

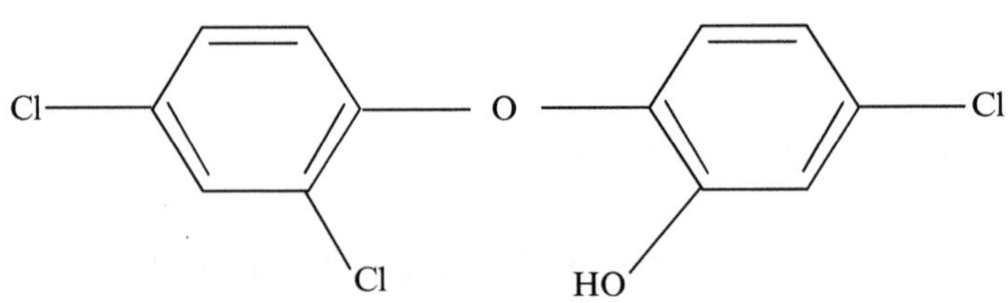

2,4,4'-Trichloro-2'-hydroxydiphenyl ether

Fig. 8.1 Structural formula of triclosan

A barbed suture, which is self-anchoring with no knots required for wound closure, is manufactured from polydioxanone. This suture consists of axially barbed segments on each side of a midpoint at which the barbs change direction. This wound closure device appears to offer gastrointestinal closure comparable to the poly(GA-co-TMC) suture.

One of the major problems in neurosurgery is neuroma formation, which often results in neuropathic pain that can be unbearable for the patient. Neuroma formation can be influenced by the material of suture fibers used in nerve repair. Inhibition of axonal outgrowth accompanied by neuroma formation appears in microsurgical nerve repair as a reaction to common microsuture materials like silk, nylon, or PGA. While nearly every suture material results in granuloma formation, giant cell invasion, and fibrosis, nylon sutures cause less foreign-body reaction than other materials, making them the gold standard in microsurgical nerve repair. Interestingly, recent findings revealed the advantages of spider silk fibers in guiding Schwann cells in nerve regeneration [2, 3]. Therefore, Kuhbier et al. attempted to braid microsutures from native spider silk fibers [4]. Spider silks have proven to possess high antimicrobial properties, which may lead to lower infection rates in peripheral nerve surgery, which is often caused by trauma and leads to septic wound conditions. Additionally, recombinant spider silk has extremely low pyrogenicity, lending further support for its usefulness in trauma surgery where sepsis is common. Both spider silk and nylon are polyamides containing amide bond (–NHCO–) in the main chain. Microsutures braided of native spider dragline silk were manufactured containing either 2×15 or 3×10 single fibers strands. The tensile strength and stress–strain ratio (SSR) are shown in Fig. 8.2. The constructed spider silk sutures had a median thickness of 25 μm, matching the USP definition of 10–0. Maximum load and tensile strength for both spider silk microsutures were significantly more than twofold higher than for nylon sutures; the SSR was 1.5-fold higher. All values except elasticity were higher in 3×10 strand sutures than in 2×15 strand sutures, but the difference was not significant. With regard to mechanical properties, these sutures were superior to nylon sutures. Since spider silk displays high biocompatibility in nerve regeneration, its usage in microsurgical nerve repair may be considered in the near future.

Numerous alternatives to sutures have been proposed because anastomoses performed with sutures can lead to failure from intimal hyperplasia and foreign body reaction, resulting in vascular occlusions, prolonged hospital stays, and mortality [5, 6]. To overcome these problems of suturing, a new technology for sutureless vascular anastomosis was reported by Chang et al. [7]. They used a thermoreversible triblock polymer for temporal maintenance in an open lumen for approximation with 2-octyl cyanoacrylate adhesive. In animal studies, this sutureless anastomosis was found to be five times faster than the traditional hand-sewn control. Moreover, even vessels with less than 1 mm, which would be too delicate for suturing, were able to be reconstructed. This approach decreased both inflammation and scarring after 2 years.

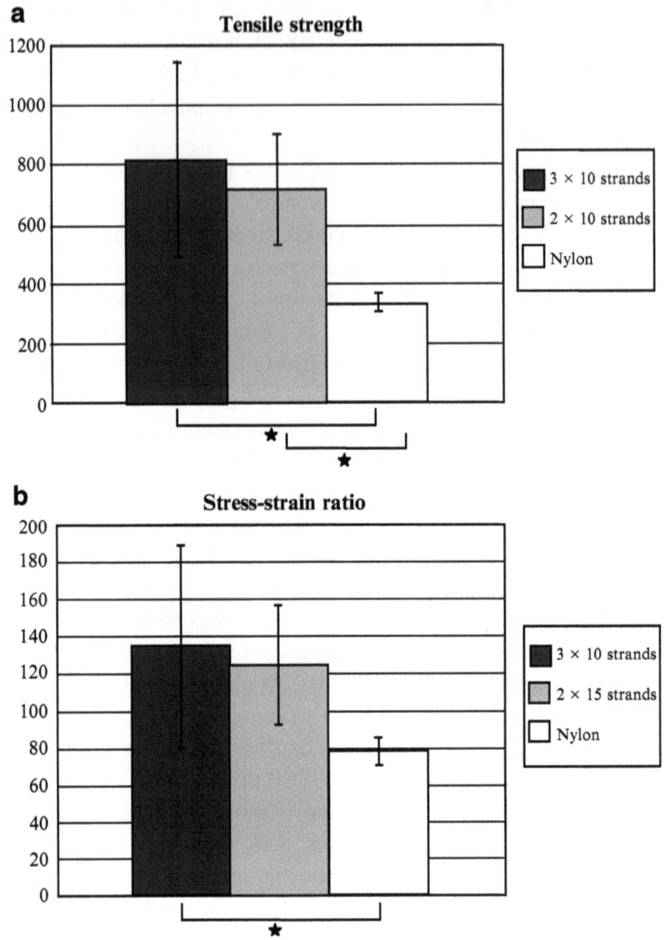

Fig. 8.2 Comparison of the mechanical properties of braided spider silk sutures and nylon sutures. (**a**) Comparison of the tensile strength of 2×15 silk strands, 3×10 silk strands, or a nylon suture. (**b**) Comparison of the stress–strain ratio (SSR) of 2×15 silk strands, 3×10 silk strands, or nylon suture. *Error bars* indicate standard deviation; *asterisks* mark significance level of $p<0.05$ [4]

References

1. Firestone DE, Lauder AJ (2010) Chemistry and mechanics of commonly used sutures and needles. J Hand Surg Am 35:486–488
2. Allmeling C, Jokuszies A, Reimers K, Kall S, Vogt PM (2006) Use of spider silk fibres as an innovative material in a biocompatible artificial nerve conduit. J Cell Mol Med 10:770–777
3. Allmeling C, Jokuszies A, Reimers K, Kall S, Choi CY, Brandes G, Kasper C, Scheper T, Guggenheim M, Vogt PM (2008) Spider silk fibres in artificial nerve constructs promote peripheral nerve regeneration. Cell Prolif 41:408–420

4. Kuhbier JW, Reimers K, Kasper C, Allmeling C, Hillmer A, Menger B, Vogt PM, Radtke C (2011) First investigation of spider silk as a braided microsurgical suture. J Biomed Mater Res B Appl Biomater 97:381–387
5. Zubilewicz T (2001) Injury in vascular surgery – the intimal hyperplastic response. Med Sci Monit 7:316–324
6. Charlson ME (2003) Clinical practice care after coronary-artery bypass surgery. N Engl J Med 348:1456–1463
7. Chang EI, Galvez MG, Glotzbach JP, Hamou CD, El-ftesi S, Rappleye CT, Sommer K-M, Rajadas J, Abilez OJ, Fuller GG, Longaker MT, Gurtner GC (2011) Vascular anastomosis using controlled phase transitions in poloxamer gels. Nat Med 17:1147–1152

Chapter 9
Conclusions

Abstract This chapter summarizes the present status of bioabsorbable biomaterials and devices used in surgery, which include surgical sealants, antiadhesion barriers, fixation devices, and sutures. These devices are generally indispensable for surgery to heal injuries and prevent tissue adhesions, but the products that are currently available, such as fibrin glue, have unsolved problems. They include low adhesion strength, difficult handling, tedious preparation, and poor balance between bioabsorption rate and mechanical strength maintenance. For instance, bioabsorbable bone fixation devices need improvement in bioabsorption kinetics. Several kinds of growth factors are commercially available to enhance wound healing, but their carriers are not always effective for their sustained release, and, in addition, their cost is very high. Recently, several techniques have been developed to concentrate patient's own platelets which contain large amounts of growth factors. This last chapter seems to suggest that a new era is just starting with regard to more effective and efficient devices for surgical operation.

Two kinds of tissue injuries—intentional and unintentional—develop during surgery. Intentional injuries include those that develop during skin cutting to secure the area for operation, resection of diseased tissues or organs, and organ transplantation or device implantation. Furthermore, tissue injuries also occur accidentally during operations presumably because of mechanical contact of metallic devices and gloves (used for surgery) with the patients' internal organs and tissues. In most of the cases, such injuries heal naturally or by some external means. Surgical application of bioabsorbable biomaterials is an external means of healing the injuries. As described in the previous chapters, hemostatic glues, surgical sealants, antiadhesion barriers, fixation devices, and sutures have tremendously contributed in healing wounds, arresting bleeding, and preventing tissue adhesion and bone movement. Since only few bioabsorbable materials have been developed, surgeons have long been expecting many developments in these supporting materials.

One of the widely recognized problems associated with bioabsorbable materials is the poor performance of and the high risk associated with fibrin glue, although it

S. Suzuki and Y. Ikada, *Biomaterials for Surgical Operation*,
DOI 10.1007/978-1-61779-570-1_9, © Springer Science+Business Media, LLC 2012

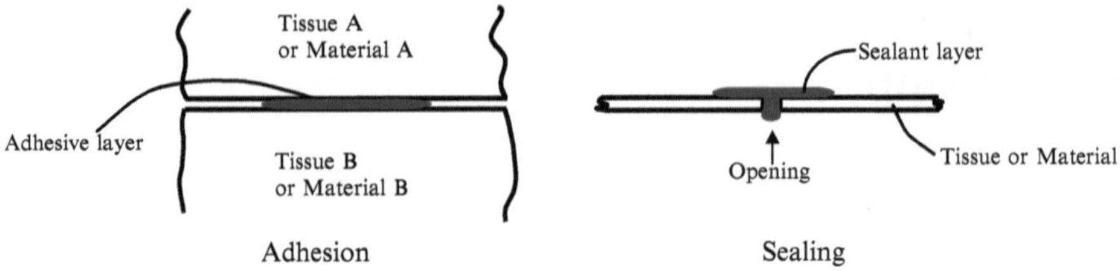

Fig. 9.1 Adhesion and sealing

is most widely used among all surgical glues (adhesives and sealants). The fibrin glue has poorer adhesive strength (for adhesion to the tissue surface) than nonmedical conventional adhesives used at home and in industries. Major reasons for the poor performance of fibrin glue may be the presence of fluids in the surrounding environment, low solution concentration of fibrinogen, low density of cross-links formed by the catalytic action of thrombin, and unsuitable characteristics of the constituting polymer chains. When the fibrinogen is harvested from the patient's own blood (in hospitals) so as to avoid pathogen contamination, the fibrinogen concentration of the prepared solution is much lower than that of commercially available fibrin glues. Hence, the fibrin glues applied to surgical wounds are generally not derived from patients' blood, even at a risk of a viral infection in the usage of commercially available fibrin glues. The fibrinogen concentration in the blood plasma is 0.2–0.4 g/100 mL, while that in commercially available fibrin glues is around 8–12 g/100 mL, which is much lower than the polymer concentrations in other conventional adhesives. To address this issue, a variety of advanced hemostats have been brought into the market, but apparently, none of them have the efficacy of fibrin glues. A trade-off between performance and biosafety of adhesives is frequently observed. Commercially available surgical adhesives other than fibrin glue include α-cyanoacrylates, gelatin-resorcinol-formaldehyde glue (GRF®), and albumin glue (Bioglue®); but all of these adhesives lead to toxicity because of the high concentration of formaldehyde or glutaraldehyde released by or present in these adhesives.

Such adhesives are often referred to by different technical terms such as "sealant." A sealant is defined as an agent that can seal openings such as apertures, cracks, fissures, gaps, holes, and slits, whereas an adhesive (and glue) is defined as an agent that can firmly bind material A to another material B or to A itself, as shown in Fig. 9.1. Both agents are in liquid state when applied to the object's surface(s), following which they solidify. This solidification leads to strong adhesion between the thin adhesive or sealant layer and the object's surface(s). Interestingly, such adhesive agents are only rarely used in surgical operations to facilitate close contact and binding between tissues. When a firm binding is required between two soft tissues, sutures are predominantly used. This is so because, when an adhesive is used to bind two tissues, an unabsorbed solidified adhesive layer remains between two tissues, thus disturbing the natural wound healing process. This phenomenon is in marked contrast with suturing, where the suture thread occupies only a very tiny fraction of the wound surface.

In the field of surgery, technical terms of adhesive and glue should be replaced with "sealant." Sealing of wounds is extremely important in an operation in order to stop bleeding. Bleeding from large blood vessels can be avoided by ligating the vessels, whereas electric cauterization should be used to ligate wounds in small blood vessels. However, these techniques cannot be used to arrest bleeding from very small blood vessels. On the contrary, sealants are very effective for such slight but continuous blood leak, unless the adhesion of the sealant layer to the affected area is insufficient. As shown in Fig. 9.1, strong adhesion is critical for both adhesives and sealants between the thin solidified layer and the surface of the material (or tissue) in contact with the agent. Since adhesive strength is an important factor in case of sealants, terms such as surgical "adhesives" and fibrin "glues" may have widely come into use as they convey the idea of sealants.

Various medical applications have been used as surgical sealants. They are mainly applied to stop the fluid or air leakage from the affected tissues or devices (like synthetic vascular graft and dura mater), which is done by sealing the wounds. The fluids that could leak include blood, pancreatic juice, bile, and cerebrospinal fluid (CSF). Air leakage occurs solely from diseased lungs. Hemostasis by mechanical (not biological) suppression of bleeding is the major application of surgical sealants. In this regard, the adhesive force between the affected tissue and the solidified sealant layer is crucial for effectively stopping the leakage. Since the moisture and the blood present on the affected tissue surface clearly reduce the adhesive strength, the tissue should be cleaned to the maximum possible extent. Further, spouting of blood from the areas adjacent to the tissue should be arrested.

In the recent future, if research and development are performed on the basis of the abovementioned considerations, the development of new, more effective, and less toxic sealants from bioabsorbable biomaterials will be possible.

Another issue that needs to be addressed for the use of bioabsorbable polymers with regard to surgeries is the identification of protective barriers against tissue adhesion that would be more effective and easy to handle than the currently available commercial protective barriers. If severe tissue adhesion occurs during an operation, the patient will show an abnormal body state accompanied by pain. Moreover, in cases involving lesions arising from tissue adhesion, the second operation will have to be prolonged. Tissue adhesion is known to occur even if surgeons perform the operations with a lot of care and caution. Furthermore, tissue adhesion takes place not only between two injured tissues but also between an injured tissue and an apparently unaffected (normal) tissue with a very small undetected injury on its surface. Even if the other tissue is actually unaffected, omental tissue may adhere to the injured tissue, necessitating urgent treatment of the injury.

If surgeons anticipate postoperative occurrence of tissue adhesion, they can place a barrier membrane at the probable adhesion site in order to avoid tissue adhesion, as illustrated in Fig. 9.2. The minimum requirements for the barrier membrane to resist tissue adhesion are sufficient tear strength in the wet state; easy handling, especially during endoscopies; a bioabsorption rate matching that of wound healing; and biocompatibility of the other open surface of the barrier membrane that is not in direct contact with the injured tissue. Moreover, the membrane should have

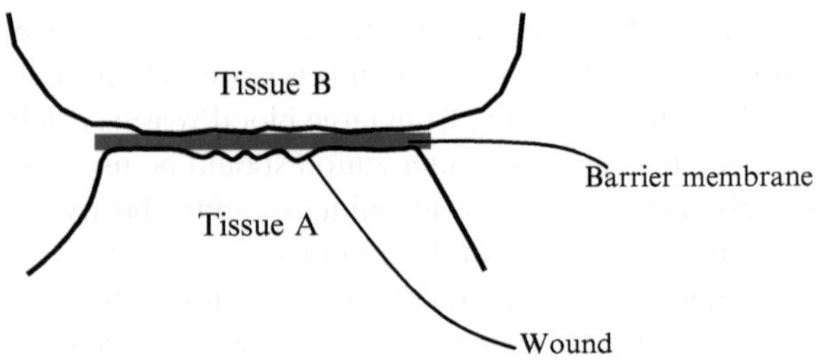

Fig. 9.2 Application of a barrier membrane to prevent adhesion between a wounded tissue A and the opposing tissue B

the ability to remain fixed without getting peeled off at the site with a high probability of tissue adhesion. Firm fixation of the barrier by using adhesives may not be required unless tissue movement around the barrier is very active. Surgeons are not always satisfied with the currently available barrier materials. On the contrary, too firm fixation of the barrier membrane might hinder the wound healing. Anyway, considering the simple function of a barrier membrane, development of membranes meeting all the expectations of surgeons should not be difficult. In the last century, even nonbioabsorbable membranes were used as barriers to prevent tissue adhesion. However, at present, membranes that elicit significant foreign-body reactions are not used for prevention of tissue adhesion.

Currently, among bone fixation devices, products based on bioabsorbable polymers have a lower market share than those derived from metals. In order to increase this share, polymer fixation devices should have improved bioabsorption-enhancing capacity with high mechanical strength. The head part of PLLA screws is the most important component in terms of mechanical strength because the screwdriver exerts a lot of force on the head. Secondary surgery is not required after using bioabsorbable devices, and metallic implants are associated with continuous risk of foreign-body reactions even after the reunion of fractured bones. Therefore, no one will prefer metallic implants to bioabsorbable devices.

Currently, sutures have the largest market share among the synthetic bioabsorbable polymer materials used in medicine. A variety of bioabsorbable polymers have been synthesized to fabricate different types of monofilament and multifilament sutures with a wide range of performances. Therefore, except in the case of specific applications, development of new types of bioabsorbable sutures is not expected from researchers or surgeons.

Future direction of research in the field of bioabsorbable biomaterials is aimed at creating new devices or systems that can enhance wound healing and eventually reduce the hospital stay of the patients. One possible approach to this may be by taking advantage of the growth factors essential for wound healing. Studies from the last 3 decades have shown that growth factors such as bFGF, VEGF, and BMP actually accelerate wound healing and tissue regeneration if they are applied to the body efficiently. However, development of an effective system for sustained delivery of growth factors remains a great challenge. In addition, in comparison with other drugs, growth factors are very expensive and difficult to obtain.

Recently, platelets have attracted much attention because they contain a number of growth factors in the α-granule. This finding is of great interest to surgeons because it suggests the possibility of harvesting growth factors in the autologous form from the blood of patients before or during an operation. However, one possible problem associated with the platelet application is the sustained release of growth factors for the required period of time. Most of the studies that have attempted to address this challenge have made use of the fibrin gel obtained from the patient during platelet collection; however, the fibrin matrix used to support growth factor release is likely to get rapidly hydrolyzed at the application site. As a result, the growth factors may be released in large amounts during the first few hours after application. In this regard, a fibrin hydrogel with higher cross-link density may be a choice as a carrier for sustained release of growth factors because the higher cross-link density lowers the bioabsorption rate of fibrin hydrogel.

Since multiple research groups are currently engaged in exploring more effective carriers for the sustained release of growth factors, we expect that new approaches to promote wound healing and tissue regeneration will be developed in the near future.

Index

S. Suzuki and Y. Ikada, *Biomaterials for Surgical Operation*,
DOI 10.1007/978-1-61779-570-1,

About the Authors

Shuko Suzuki obtained her PhD degree in polymer chemistry from Queensland University of Technology, Australia. After JSPS fellowship at Nara Medical University, she joined GUNZE Ltd., Japan, and is currently working as research fellow at Nara Medical University.

Yoshito Ikada has over 40 years of experience as a professor at a number of universities in Japan and China in the field of biomedicine. He is currently at Nara Medical University, and he is also emeritus professor at Kyoto University and guest professor at Peking University. He has received the Prize of Japanese Society for Biomaterials and was President of the Japanese Society for Biomaterials for 4 years. He was also the Chairman of National project, "Tissue Engineering," by Japan Society for the Promotion of Science. He authored "Tissue Engineering: Fundamentals and Applications" (Academic Press, 2006).

About the Authors

MIX
Papier aus verantwortungsvollen Quellen
Paper from responsible sources
FSC® C105338

If you have any concerns about our products,
you can contact us on
ProductSafety@springernature.com

In case Publisher is established outside the EU,
the EU authorized representative is:
Springer Nature Customer Service Center GmbH
Europaplatz 3, 69115 Heidelberg, Germany

Printed by Libri Plureos GmbH
in Hamburg, Germany